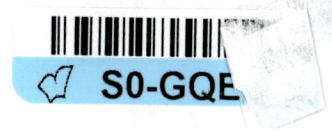

DOING STATISTICS

With

MINITAB for Windows™
Release 10

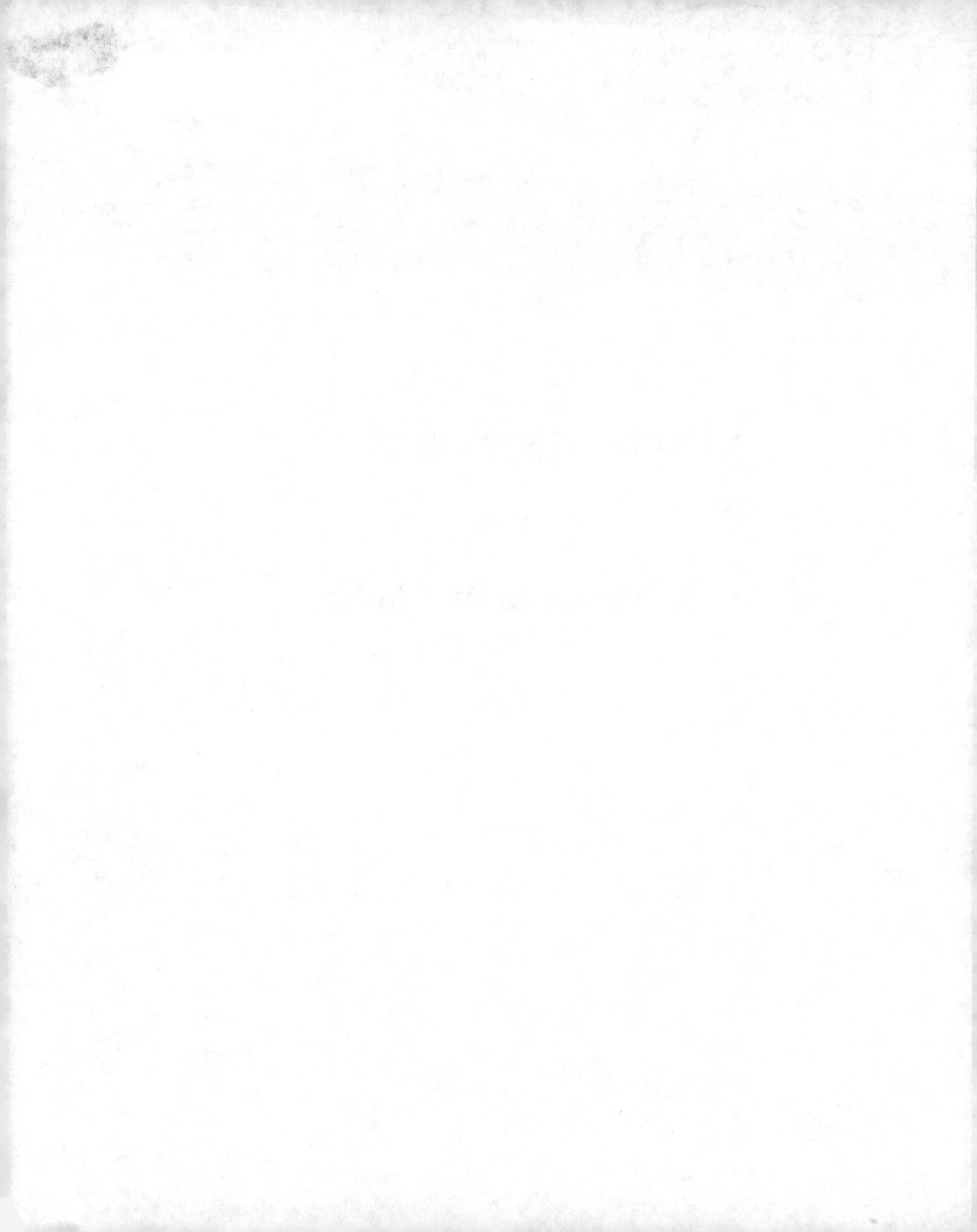

DOING STATISTICS

With

MINITAB for Windows™
Release 10

AN INTRODUCTORY COURSE SUPPLEMENT
FOR EXPLORATIONS IN DATA ANALYSIS

Marilyn K. Pelosi, Ph.D
Western New England College

Theresa M. Sandifer, Ph.D
Southern Connecticut State University

John Wiley & Sons, Inc.
New York • Chichester • Brisbane • Toronto • Singapore

ISBN 0-471-30471-9

Printed in the United States of America

10 9 8 7 6 5 4 3 2

This book is dedicated to friendship.

"It is not often that someone comes along who is a true friend and a good writer. "

- E.B. White, <u>Charlotte's Web</u>

Table of Contents

Preface

To The Student:

Data! Data! Data! It is everywhere and growing at an incredible rate. As our computer technology allows us to capture, store and retrieve more and more data, we must be able to transform that data into information. The information can then be used to make business decisions at all levels ranging from day-to-day decisions to short-term decisions to long-term strategic planning decisions. We can no longer compete internationally by making decisions based solely on experience and intuition. We must be able to see the information in the data.

This workbook is designed to help you learn how to see the valuable information in data. Each chapter is designed to allow you to explore a single statistical concept by investigating a data set. The situations which led to the collection of the data sets are based on actual consulting experiences of the authors.

It becomes quickly obvious that you can not investigate a data set without the assistance of a computer package. The computer package which is used in this workbook is known as Minitab for Windows Release 10. The commands necessary to use this package are explained in each chapter. If you are using the Student Edition of Minitab for Windows there may be some minor differences between the screens you see and the ones shown in the workbook.

To The Instructor:

This workbook is designed to be used by students enrolled in a basic course in Statistics. It is built around the premise that students must *Do Statistics* in order to truly understand the statistical concepts. Thus, each chapter reinforces a topic from a standard statistics course. It is expected that the student can work independently or as a member of a team of students.

The workbook is self-contained in that all of the instructions needed to use the software are explained in the appropriate chapter. NO additional material needs to be presented by the instructor. The instructor can assign the workbook chapters which go with the statistical lecture material. Space is provided for the students to write the answers in the workbook, if so desired.

Each chapter contains two types of exercises: (1) exercises to help the student learn the command structure of Minitab for Windows Release 10 and (2) exercises to help the students see the statistical concepts in action. The chapter is motivated by an actual problem drawn from the authors consulting experiences. The dataset is then explained and the remainder of each chapter is dedicated to two major goals:

(1) to provide the student with detailed command instructions for using Minitab for Windows to accomplish the desired statistical analysis;

and
(2) to provide the student with a set of structured exercises designed to direct the students statistical thinking in order to reinforce the statistical tool of the chapter.

Acknowledgment

The authors wish to acknowledge Minitab Inc. for supplying each of them with a copy of Minitab for Windows Release 10 and a copy of the Student Version of Minitab for Windows. Further information about the software can be obtained by contacting Minitab directly at the following address:

Minitab Inc.
3081 Enterprise Drive
State College, PA 16801-3008 USA

Chapter 1 "Why *Do* Statistics in Minitab for Windows?"

An Introduction to the Workbook

Section 1.1 Overview

This chapter discusses the following topics:

- Objectives of the workbook
- Description of the Data Sets
- Organization of the Workbook
- How to use the Workbook

Section 1.2 Objectives of the Workbook

This workbook has two major objectives. They are:

(1) to allow you to *see* the statistical concepts taught in a standard course in statistics at work by *doing statistics* with real life data sets using the Minitab for Windows software and;

(2) to help you *learn how to use Minitab for Windows*.

At first glance, these may appear to be rather separate and unrelated objectives. In fact, they are supportive of each other. The workbook is designed around the premise that in order to truly understand the statistical concepts you must *do statistics* using real life data sets and in order to *do statistics* you must use a statistical software package. Thus by exploring the data sets you will learn both the statistical tools and the command structure of the software.

Section 1.3 Description of the Data Sets

Each chapter uses one data set whose subject is hinted at in the title of the chapter. In a few cases a data set may be used for more than one chapter. In these cases, the details of the data set are provided in the first chapter that uses this data set.

The data sets are based on actual business situations and in all cases reflect actual problems. They are drawn from the consulting experiences of the authors. The names of the companies have not been provided in order to maintain confidentiality agreements. However, the problem descriptions, the variables and the data are real. The data sets are also large enough to (1) allow you to see the need for a statistical software package and (2) to explore patterns and relationships.

Section 1.4 Organization of the Workbook

The workbook is divided into two major parts.

Part I: Basics of Windows and Minitab for Windows

Chapter 2 explains some basic Windows commands. Chapter 3 gets you started in Minitab for Windows Release 10. It covers the basic structure and commands for Minitab for Windows Release 10. After reading this chapter you will be ready to move on to the remaining chapters of this workbook. Specific commands and software features will be explained in the chapters where they are needed for the statistical analysis.

Part II: Statistics

Chapters 4-15 are the chapters where you will be *Doing Statistics*.

Each chapter is designed to allow you to explore a single statistical concept by investigating a data set. The chapters follow the normal sequence of topics in an introductory statistics course. Each chapter has the following structure:

- Summary of Statistical Objectives
- Problem Statement
    The problem statement section of each chapter describes the business situation which led to the collection of the data.

- Characteristics of the data set
  This section of each chapter explains the details of the particular data set including information such as the filename, the name and column location of the variables, size of the file, etc.

- Tutorials
  The next few sections of each chapter give the instructions for how to use Minitab to accomplish the particular statistical concept. Examples are provided using the data set for that chapter.

- Investigative Exercises
  These exercises are designed to allow you to analyze the data set using the specific tools explained in the chapter. There is space provided for you to either paste in printed output from Minitab or to write in the answers by hand based on the screen display from Minitab.

  The exercises are highly structured in the early chapters leading you to a directed analysis of the data set. You will always be asked to draw conclusions and make recommendations on the basis of your analysis.
  As the chapters progress, it is expected that you will become increasingly familiar with the sorts of questions that should be asked, the types of calculations which might be informative and the types of comparisons which should be examined.

Section 1.5 How to Use the Workbook

This workbook is designed to be used by the students taking an introductory course in statistics. It is assumed that the student has been presented the traditional lecture material for each of the concepts prior to working with the corresponding chapter of the workbook.

The workbook chapters could be assigned as homework on an individual basis or to teams of students. Alternatively, the workbook chapters could be used during class lab sessions.

Chapter 2 The Basics of Windows

Section 2.1 Overview

This chapter will review some of the basic features of the Windows Operating System. It is not designed to be a comprehensive introduction to Windows and only covers those features of Windows that are essential for using Minitab for Windows and this workbook. It assumes that you are familiar with the use of a mouse and that Windows is installed and *currently running* on the computer you are using.

Section 2.2 The Windows Desktop

When Windows is open the screen you see is called the *desktop*. This is where the tools needed for using Windows and running your applications (software packages) are located. Each tool or application is represented by a small symbol called an *icon*. The applications and icons are grouped together in ***program groups***. Each program group opens up as a window.

The main portion of the desktop is the ***Program Manager***. Usually, when Windows is running the Program Manager is open and the Program Manager *window* is displayed. A *window* is a rectangular area of the screen usually dedicated to a particular set of tasks. The Program Manager window contains all of the basic tools for Windows as well as the icons for any applications that are installed on the system. Figure 2.1 shows the Program Manager window with another window (the **Main** window) open within it. The terms that have been defined as well as several others are illustrated in the figure.

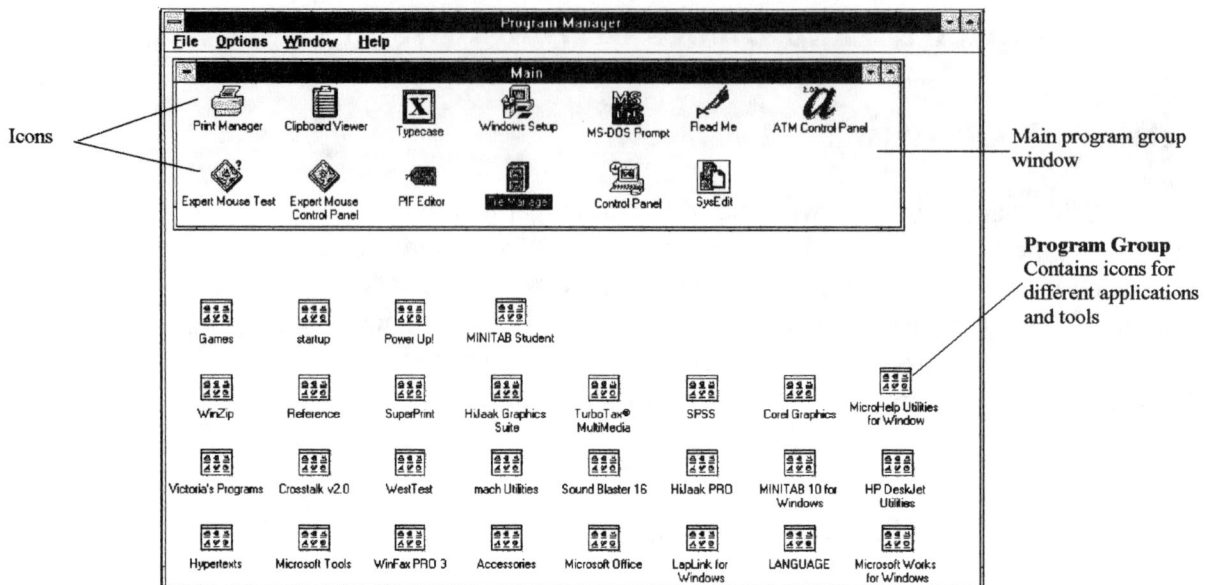

Figure 2.1 The Windows Desktop

Depending on the system you use, your desktop may not look exactly like the one shown. It may have fewer or different program groups and different windows might be opened. Generally the setup should be similar to the one in Figure 2.1.

Section 2.3 Parts of a Window

In order to use Windows effectively there are certain features that you should know about. A basic window is shown in Figure 2.2. The important features are illustrated and explained in the figure. To learn more about these features consult your Windows documentation.

Control-menu box
Used to resize, move, minimize and maximize a window

Menu Bar
Allows you to execute commands for the application or tool of the current window

Window Title Bar

Minimize and Maximize Arrows.
Allows you to change the size of the current window

Vertical and Horizontal Scroll Bars
Allows you to move up and down and left and right in the current window so that you can view hidden portions of the screen

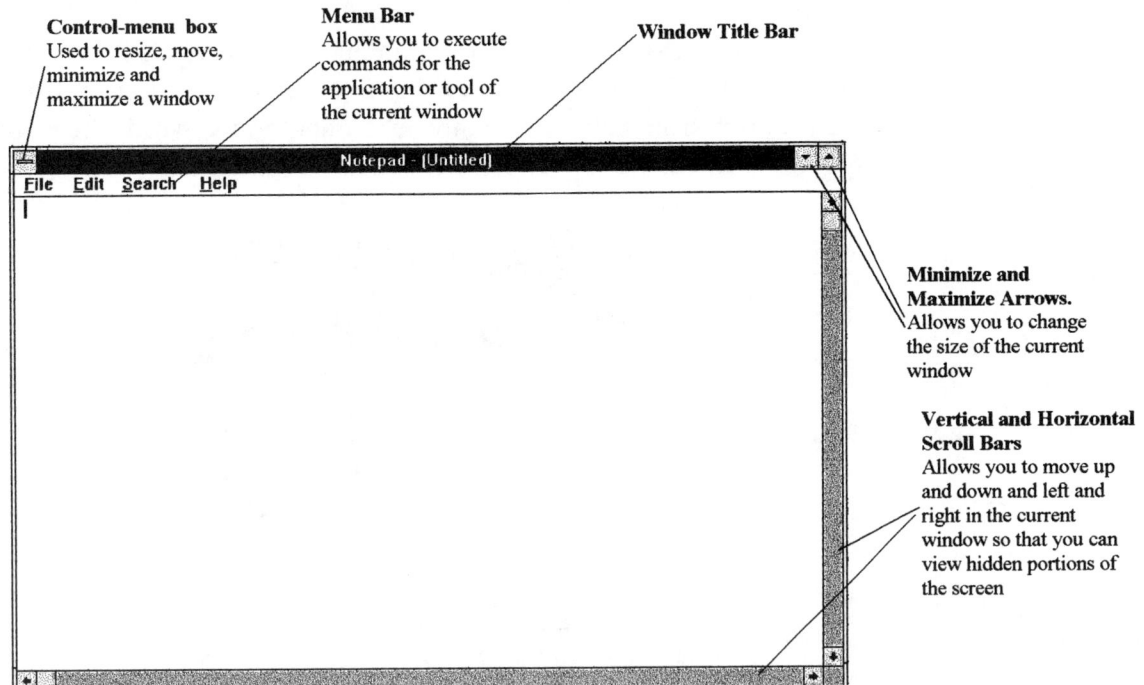

Figure 2.2 Parts of a Window

It is possible to have more than one window open at a single time. The *active window* is the one you are currently working in. Its title bar is typically a different color from the title bars of the other open window. To make another window the active one, you can either click on any portion of that window or use the **Window** menu from the main application tool bar.

Section 2.4 Other Windows Features

This section of the chapter will illustrate some of the other Windows features that you will encounter in Minitab.

Section 2.4.1 The Menu

Every application or tool in Windows has a *menu bar*. You use this menu bar to execute commands for the application or tool you are running. To access a menu you can either click on the menu title from the menu bar or use a combination of the ⁅ALT⁆ key and a letter. The letter you use is the one that is underlined in the menu choice.

When you access a menu, a list of commands available from that menu opens from the menu bar. For example to open the **File** menu from the program manager, use the mouse to click on the word **File** or hit 🖮 🖮. The menu that appears is shown in Figure 2.3.

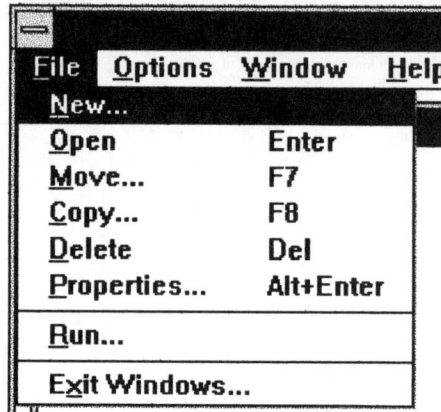

File	Options	Window	Help
New...			
Open	Enter		
Move...	F7		
Copy...	F8		
Delete	Del		
Properties...	Alt+Enter		
Run...			
Exit Windows...			

Figure 2.3 The File Menu

To select an item from the menu, you can
- use the mouse to click on the item
- use the 🖮 to move to the item and press 🖮
- type the letter that is underlined for that item.

Section 2.4.2 The Dialog Box

Often when you select an item from a menu, a *dialog box* will open. The dialog box allows you to specify the way that the command will be executed. Dialog boxes are an important feature of Minitab so it is important that you are familiar with some of the basic features. Not all dialog boxes contain the same features but they have many common features. The dialog box shown in Figure 2.4 is a typical dialog box in Minitab. The important features are illustrated and explained in the figure.

Option Button
Clicking on one of
these buttons allows
you to select one of
several *mutually
exclusive* options.

Text Box
Allows you to enter
text directly by typing
or by selecting from
an associated list

List Box
Presents a list of
available choices
when the cursor is
positioned in a text

Check Box
Clicking on one of
these boxes allows
you to select or clear
the option

Drop-down list box
When you click on the
arrow a list of
available choices is
displayed.

Command Button
Clicking on these
buttons executes the
command displayed
or opens another
dialog box.

2-Sample t

◉ Samples in o*n*e column

Samples:

Subscripts:

○ Samples in *d*ifferent columns

First

Second

A*l*ternative: not equal to

less than

not equal to

greater than

Confidence le

□ Assume *e*qual variances

Se*l*ect

? TWOT

OK   Cancel

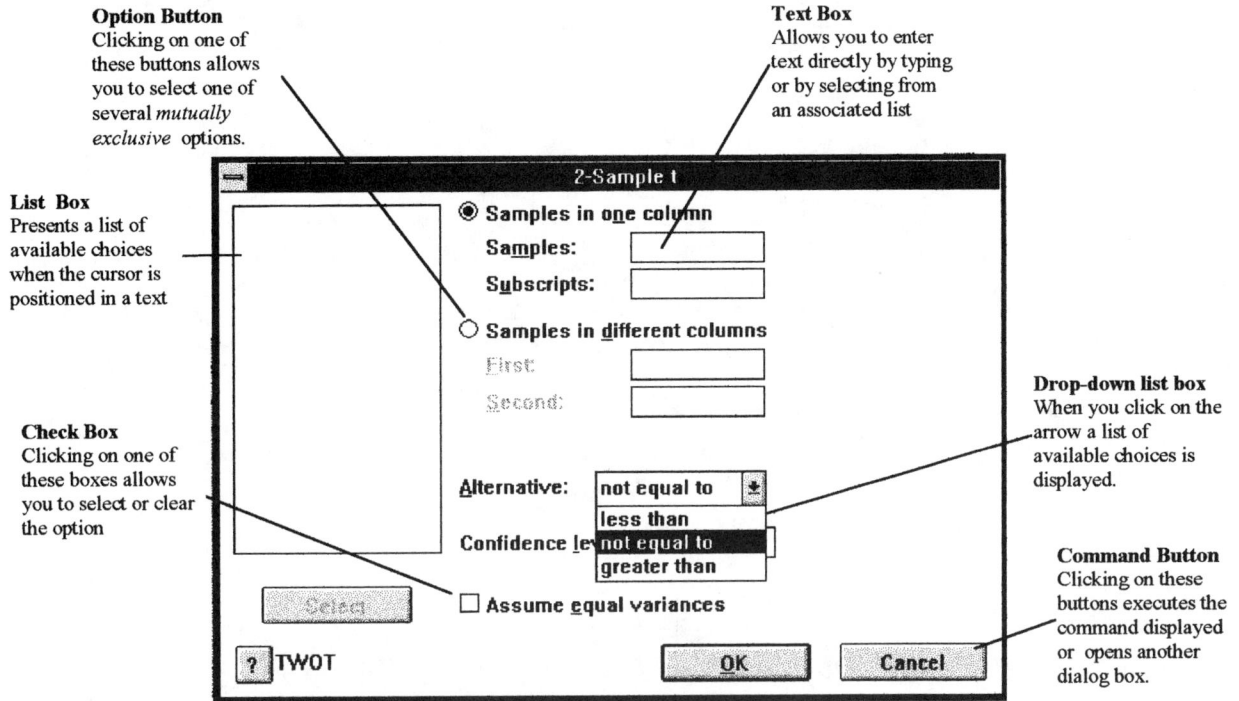

Figure 2.4 A Typical Dialog Box

To move among items in a dialog box you can either use the
mouse to click on the item you want or use [TAB] to move sequentially
to the next item in the box.

Chapter 3 The Basics in Minitab for Windows

Section 3.1 Overview

This chapter covers the basic structure and commands of Minitab for Windows Release 10. After reading this chapter you will be ready to move on to the remaining chapters of this workbook. Specific commands and software features will be explained in the chapters where they are needed for the statistical analysis.

There are some more advanced statistical features of the software that are not covered in any of the chapters in this workbook since they are not typically covered in an introductory statistics course. After you have completed the workbook you should be familiar enough with Minitab that you can use the features if you know the statistical theory. Minitab has extensive on-line help to provide you with more specific information about a statistical feature.

If you are familiar with previous non window versions of Minitab you will need to read this chapter since many things are different. If you have used a previous version of Minitab for Windows, you can skim this chapter and take note of the differences.

Section 3.2 Starting Minitab

To start Minitab, locate the Minitab program group and the Minitab icon.

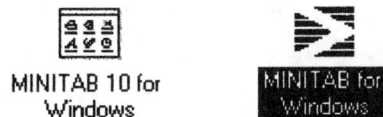

MINITAB 10 for Windows

MINITAB for Windows

Figure 3.1 Mintab Program Group and Minitab Icon

If you are using a network or a computer on which Minitab has been installed by someone else you may have to search to find the icon.

Double-click on the Mintab icon to start the program. The main Minitab window has four subwindows, the Session window, the Data window, the Info window and the History window. The windows are shown and described in Figure 3.2.

Session Window.
Displays Minitab
commands and non-
graphical output.

Main Menu Bar

Info Window.
Summarizes the
worksheet

History Window.
Records all
commands from a
session.

Data Window.
Displays data. Allows
you to enter, delete
and edit data.

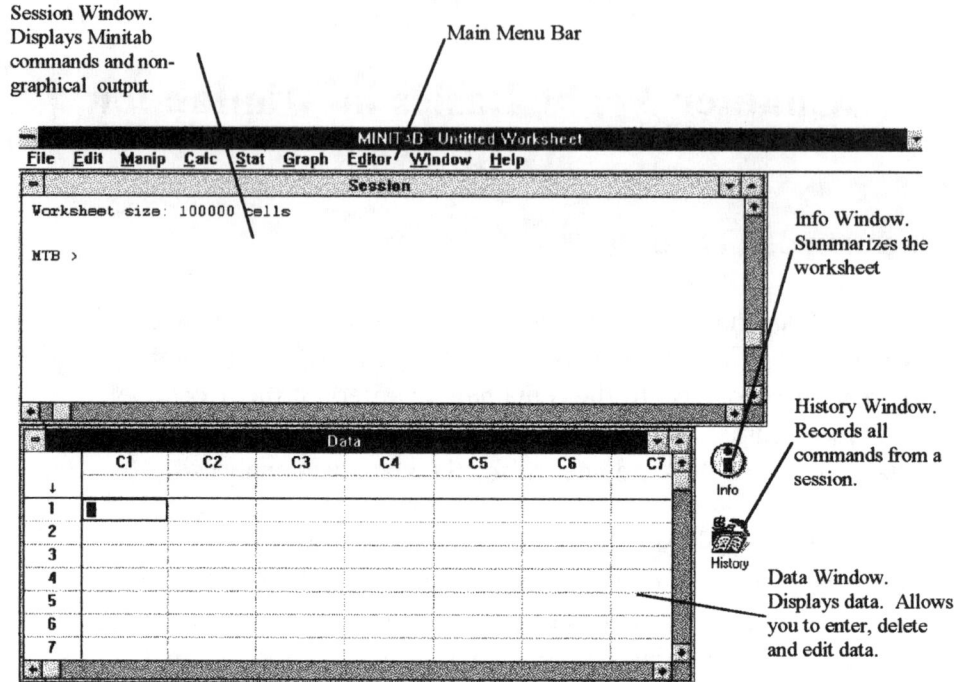

Figure 3.2 Main Minitab Window

The Session and Data windows are important and should probably always be open. The Info and History windows are fairly specialized and can be minimized for most Minitab sessions. If you do not like the default position and size of the subwindows you can alter them in any way that you like using the normal Windows methods explained in Chapter 2 or your Windows manual.

Section 3.3 The Main Menu

Before you look at any specific features of Minitab it is worthwhile to take some time examining the main menu bar illustrated in Figure 3.2. Some of the selections are common to most Windows applications, while others are dedicated to Minitab.

Two of the most often used menu commands are **File and Edit**. These menu selections deal with general features of the Minitab software. The **File** menu contains commands related to opening and saving files of different kinds. The **Edit** menu contains commands that allow you to edit the contents of the various windows.

The next five menu items, **Manip**, **Calc**, **Stat**, **Graph**, and **Editor** contain commands specific to Minitab and to statistical analysis. You will look at these menus in detail in the rest of the chapters of the workbook.

The last two menu items, **Window** and **Help** are similar to those found in most Windows applications. **Window** lets you specify how your windows appear on the screen and to switch among open windows. **Help** accesses the on-line Minitab help system.

Section 3.4 Data In Minitab

Section 3.4.1 The Data Window

A Minitab worksheet consists of *columns, constants,* and *matrices.* The largest part of a worksheet is shown in the Data window which displays the *columns* of the worksheet. Most of the time the data you want to analyze will be in column form and that is what we will concentrate on in this section.

Minitab refers to each entry in the data window by a column and a row. The columns, referred to as C#, where # is an integer value, usually represent the *Variables* and the rows usually represent the observation number for the corresponding variable. You can also refer to the columns by *Name*. Figure 3.3 identifies each of the features of the Data window.

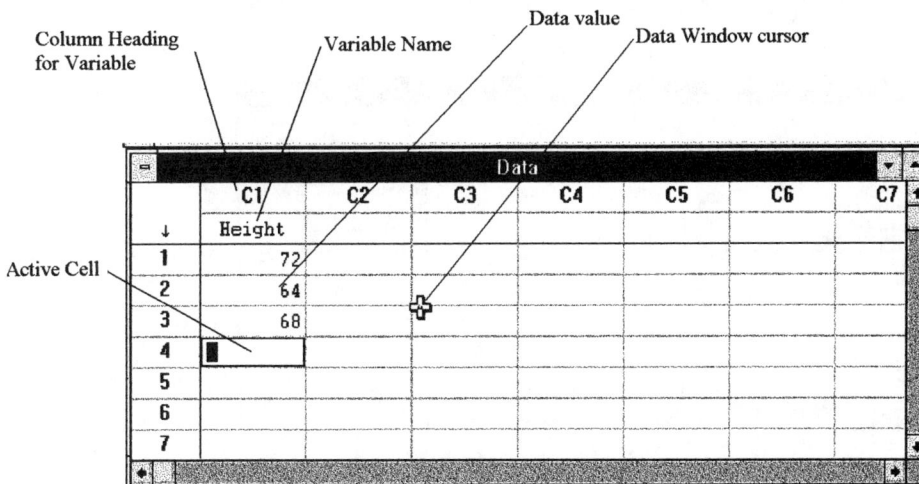

Figure 3.3 The Data Window

Section 3.4.2 Entering Data Into a Worksheet

There are several ways to enter data into a Minitab worksheet. Most of the time in this workbook you will **read data from a file** but you might want to **type in data** or **copy and paste data**.

Entering Data From a File

To enter data from a file select **File** from the main menu bar. A portion of the menu that opens in shown in Figure 3.4.

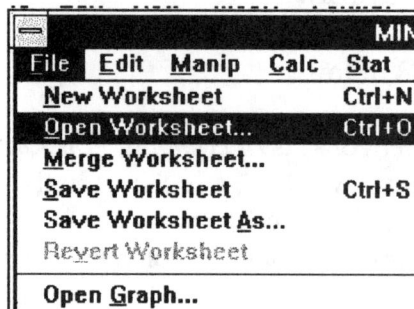

```
┌─────────────────────────────────── MIN
 File  Edit   Manip   Calc   Stat
 New Worksheet              Ctrl+N
 Open Worksheet...          Ctrl+O
 Merge Worksheet...
 Save Worksheet             Ctrl+S
 Save Worksheet As...
 Revert Worksheet
 ───────────────────────────────────
 Open Graph...
```

Figure 3.4 The File Menu

To open a new file, select **Open Worksheet** from the menu. This will **replace** the data in the current worksheet with the data in the file, that is, the current worksheet data will be lost unless it has been saved. When you select **Open Worksheet** a dialog box will open. This dialog box is shown and illustrated in Figure 3.5.

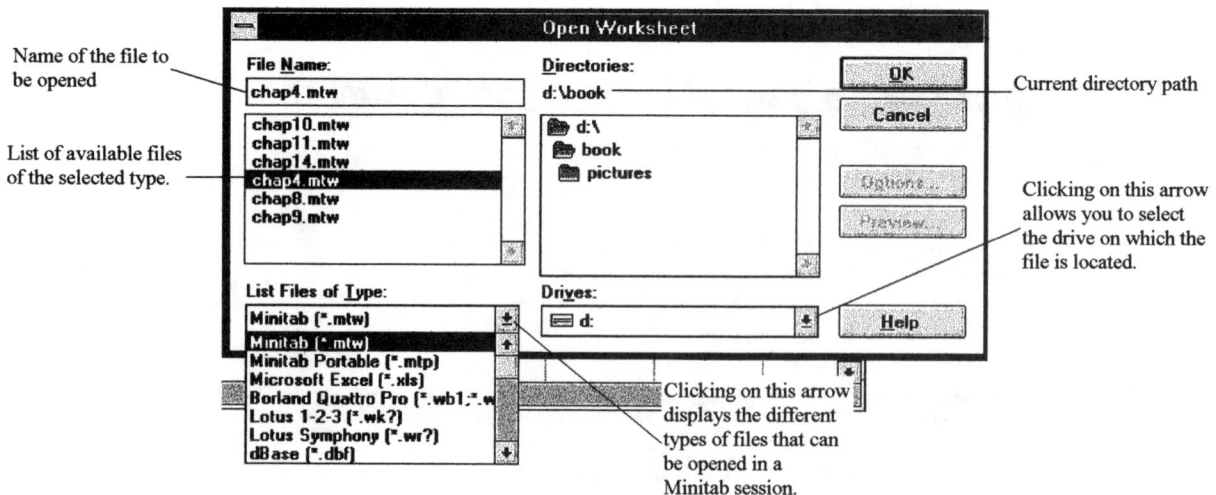

Figure 3.5 The Open Worksheet Dialog Box

Minitab allows you to open files from many well-known software packages in addition to Minitab files. Once you have chosen the location and type of the file you want to open, select the name of the file from the list and click on **OK** to open the file. Minitab worksheet file use the suffix .mtw and Minitab portable files use the suffix .mtp.

Typing Data into the Worksheet

It is possible to type data directly into the Minitab worksheet. You might want to do this for a small data set or to add data to an existing worksheet. To type data into the worksheet directly, select the column you want to use and position the cursor in the cell where you want the data to appear. Then simply type in the data value. To move to the next cell use the ⬐ or press ⏎.

Copying and Pasting Data
Sometimes in a statistical analysis you might want to look at only a portion of the data. In Minitab you can copy a portion of the worksheet to another location so that you can perform the analysis. First you must highlight the selection. Position the cursor in the upper left corner of the selection and drag the cursor to the lower right corner. The selection will be highlighted as shown in Figure 3.6.

	C1	C2	C3	C4
↓	DAY	MDSTRENG	CDSTRENG	
1	1	1006	422	
2	1	994	448	
3	1	1032	423	
4	1	875	435	
5	1	1043	445	
6	1	962	464	
7	1	973	472	

Figure 3.6 Highlighted Data Selection

The next step it to copy the selection. To do this select **Edit** > **Copy cells** from the main menu bar or use the keystrokes CTRL C. Move the cursor to the upper left corner of the new location and select

Edit > **Paste/Insert cells** or hit [CTRL] [V]. The selection will be copied to the new location.

You can also use this technique to copy data from the Windows clipboard or other applications into a Minitab worksheet.

Section 3.5 Naming Variables in Minitab

Note: When you name columns (variables) Minitab uses those names in both text and graphical output, so make sure that they are meaningful.

Since columns are used for different variables in a dataset, it would be nice to refer to each column (variable) by a *name* rather than by the C# designation. To name a variable in Minitab, position the cursor in the box at the top of the column below the C# label and type in the name you want to assign. Names can be up to 8 characters long with some restrictions on the characters. You cannot start or end a name with a blank and you cannot use the single quote (') or number sign (#) in the name. There are some other restrictions, but in general if you refrain from using characters other than letters and digits you will not have problems. An example of a named variable was shown in Figure 3.3.

Section 3.6 Saving Your Work in Minitab for Windows

There are three basic parts to a Minitab session, the **worksheet**, the **Session window** and the **graphs.** Remember that the **worksheet** consists of columns, constants and matrices. The **Session window** contains the Minitab commands and all of the text output generated by the commands. **Graphs** are separate windows generated when high resolutions graphical commands are used. Each of these parts MUST BE SAVED SEPARATELY. This section describe the method for saving worksheets and **Session** windows. Saving graphs is covered in the workbook when you create your first graph.

Section 3.6.1 Saving Your Worksheet

To save a Minitab worksheet for the first time, select **File** > **Save Worksheet As...** from the main menu. The dialog box shown in Figure 3.7 will open. This dialog box similar to the one you saw when you learned how to open a file.

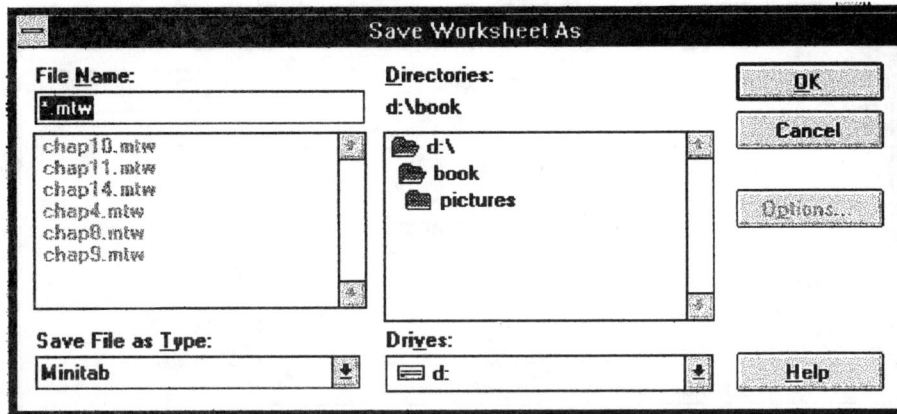

Figure 3.7 The Save Worksheet As Dialog Box

After you designate the correct drive and path for saving the file, position the cursor in the box labeled **File Name:** and type in a name for the file. Minitab uses the same file naming conventions as Windows, so you are restricted to eight characters. Minitab worksheets are automatically given the suffix, .mtw. Click on **OK** and the file will be saved.

After a worksheet has been named and saved, you can save later versions of the same worksheet by selecting **Save Worksheet...** from the **File** menu instead of **Save Worksheet As...** You will not have to go through the action of naming the file, but this will **replace** previous versions.

Section 3.6.2 Saving the Session Window

Much of the output from Minitab statistical analyses will appear in the **Session window**. If you want to examine this output at a later time or use the output as part of a word processing document you will want to save the contents of the **Session window.** To do this, first click on the title bar of the **Session** window to make it the active window. Select **File > Save Window as...** from the main menu bar and the dialog box shown in Figure 3.8 will open.

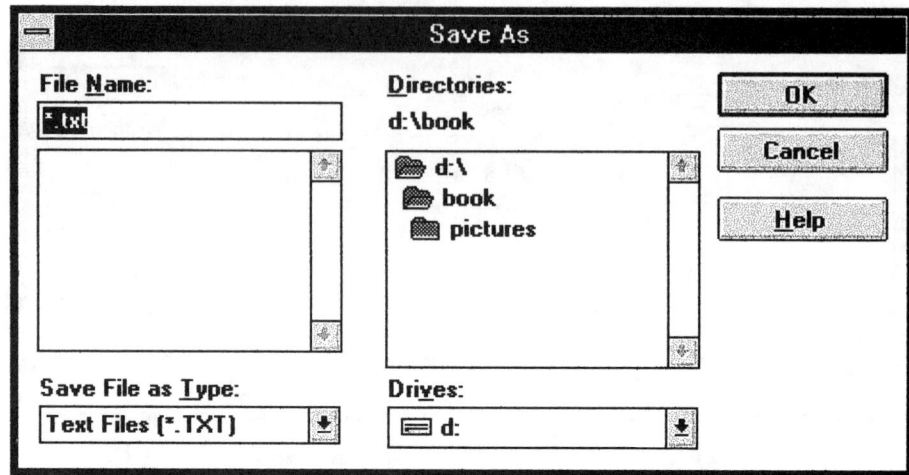

Figure 3.8 The Save Window As Dialog Box

Designate the drive and path to be used when saving the file and type in a file name in the appropriate box. Minitab appends the .txt suffix to the **Session** window file so that it can be opened in almost any word processing program.

*Note: If, during a Minitab session, you open another worksheet you will **not** be prompted to save the session window contents, so make sure you do that if you need them.*

Anytime you close Minitab you are prompted to save the *changes* to the **Session** window. If you choose to save them you will be prompted for a name for the session. Selecting an old session name will append the contents of the **Session** window to the existing file, so you will not lose the old contents of the file!

Section 3.7 Exiting Minitab

To end a Minitab session and exit from Minitab select **File** > **Exit** from the main menu bar. You will be asked if you want to save the changes to the **Session** window. If you select **Yes** you will have to name the output file as described in the previous section. If you select **No** you will exit from Minitab and return to the Windows Program Manager.

Chapter 4 "Complaint Department"

Displaying Qualitative Data

Section 4.1 Overview

Statistical Objectives: After reading this chapter and doing the exercises a student will:
- Know the function of a bar chart, clustered bar chart and stacked bar chart..
- Know the function of a pie chart.
- Be able to decide when a pie chart or a bar chart is more appropriate.
- Know when a clustered bar chart is better than multiple charts.
- Know how to examine bar charts for trends and patterns.
- Know how to draw conclusions from the information obtained from graphical displays of qualitative data.

Section 4.2 Problem Statement

Most large companies that manufacture consumer goods receive feedback from the customers. A large manufacturer of paper goods keeps track of consumer complaints for its facial tissue product line. The company receives these complaints through a toll free number that appears on the product. The data taken consists of a transcription of the actual complaint language and a classification of the complaint into a specific category. When the company gets a complaint they generally ask for additional information about the specific package. They then use the information to relate the complaints to manufacturing data so that in the future similar problems can be prevented.

Table 4.1 lists the categories and sub categories used by the company to classify the complaints:

Category	Sub category
Dispensing	Sheets Tear on Removal
	Reach In/Fallback
Foreign Material	Lint/Dust
	Other
Odor	
Miscounts	
Packaging	Defective
	Misleading
	Damaged
	Other

Table 4.1 Complaint Categories

For example, if a consumer calls the company and says "...the box of tissues I bought was not full...," the customer service representative will classify this as a miscount. Sometimes complaints can be tricky to classify. If a customer calls, for example, after a product change and says "... I eat tissues and I prefer the taste of the old <product>..." into what category should they place the complaint?

Section 4.3 Characteristics of the Data Set

FILENAME: CHAP4.MTW
SIZE: COLUMNS   5
                      ROWS   361
The first 7 lines of the data file are shown in Figure 4.1.

	C1-A	C2	C3-A	C4-A	C5
↓	MONTH	YEAR	CATEGORY	SUBCATEG	NUMBER
1	JANUARY	1989	DISPENSING	FALLBACK	80
2	JANUARY	1989		SHEETS_TEAR	118
3	JANUARY	1989	FOREIGN MATL.	LINT/DUST	53
4	JANUARY	1989		OTHER	1
5	JANUARY	1989	MISCOUNTS	MISCOUNTS	37
6	JANUARY	1989	ODOR	ODOR	5
7	JANUARY	1989	PACKAGING	ADVERTISING	0

Figure 4.1 The Complaint Datafile

Open the data file using the commands described in Chapter 3. For the exercises in this book it will be best if you **IMMEDIATELY** create a working version of the data file by resaving the file with a slightly different name. This way, if anything goes wrong (and although we *try* to avoid such problems, they still happen) you will have the original data file intact.

Section 4.4 Creating a Graph in Minitab

From the main menu bar select **Graph**. The menu shown in Figure 4.2 will open.

Remember! To open a menu click on the menu item or type ALT and the underlined letter.

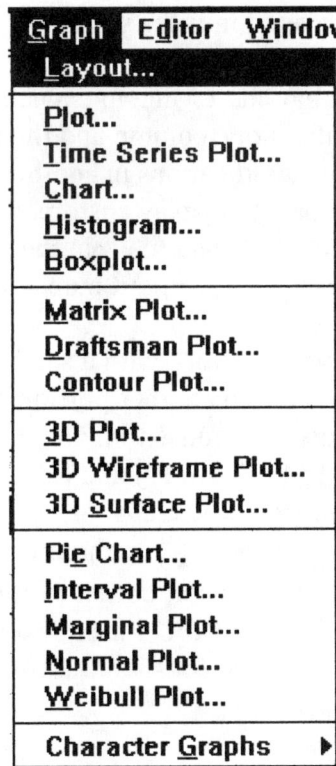

Graph   Editor   Window
Layout...
Plot...
Time Series Plot...
Chart...
Histogram...
Boxplot...
Matrix Plot...
Draftsman Plot...
Contour Plot...
3D Plot...
3D Wi**r**eframe Plot...
3D **S**urface Plot...
Pi**e** Chart...
Interval Plot...
M**a**rginal Plot...
Normal Plot...
Weibull Plot...
Character **G**raphs ▶

Figure 4.2 The Graph Menu

Most of the graphs used to display quantitative data are created using the **Chart** item from this menu, while others, such as pie charts have their own selection.

When you select one of the items from the **Graph** menu, a dialog box for creating that graph type will open. Each dialog box has items specific to the type of graph chosen, but most of the graph options contain aspects of graphing data that are common to all graphs. As you create a graph you will you will also learn about those formatting items that are common to most graphical displays of data such as titles.,

Section 4.5 How to Create Bar Charts in Minitab

Section 4.5.1 Creating the Chart

This section focuses on the basics of creating bar charts using Minitab. Examples of some of the more general features of graphs such as titles are also covered in these sections.

When creating a bar chart Minitab expects to find the classifications for the data in one column and the raw data or frequency counts for the classifications in another column. We will begin by creating a bar chart to display customer complaints for January 1989.   The data in the data file is in the correct format for Minitab to produce such a graph without much difficulty.

Since you only want to work with a section of the data file you will need to copy that section to another portion of the Minitab worksheet. To do this use the mouse to highlight the January 1989 data (From C1 Row 10 to C5 Row 10) and select **Edit** from the menu bar. From this menu choose **Copy Cells** or, if you like to work with the shortcut keys, you can hit [CTRL] [C] without opening the menu. Next, move the cursor to the top left corner of the worksheet location for the data, **C6 Row 1.** Select **Edit>Paste/Insert Cells** or use the shortcut, [CTRL] [V] . Your worksheet should look like the one in Figure 4.3.

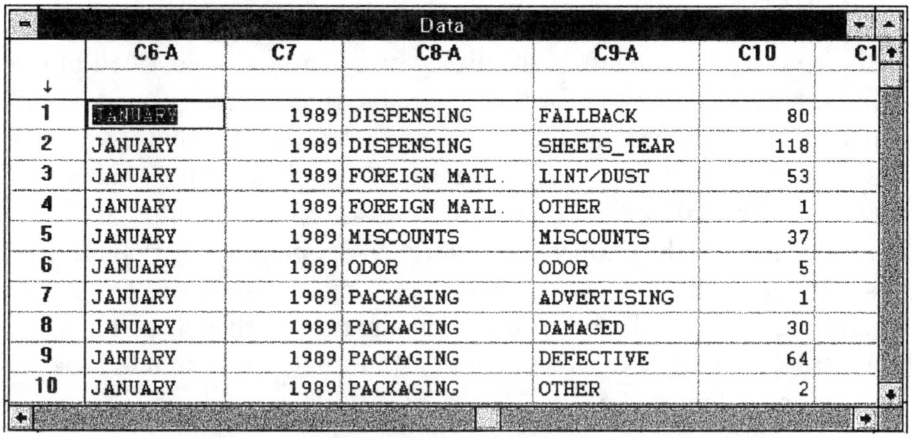

	C6-A	C7	C8-A	C9-A	C10	C1
1	JANUARY	1989	DISPENSING	FALLBACK	80	
2	JANUARY	1989	DISPENSING	SHEETS_TEAR	118	
3	JANUARY	1989	FOREIGN MATL.	LINT/DUST	53	
4	JANUARY	1989	FOREIGN MATL.	OTHER	1	
5	JANUARY	1989	MISCOUNTS	MISCOUNTS	37	
6	JANUARY	1989	ODOR	ODOR	5	
7	JANUARY	1989	PACKAGING	ADVERTISING	1	
8	JANUARY	1989	PACKAGING	DAMAGED	30	
9	JANUARY	1989	PACKAGING	DEFECTIVE	64	
10	JANUARY	1989	PACKAGING	OTHER	2	

Figure 4.3 Worksheet after Copying Cells

Now you are ready to create the bar chart. Select **Graph** from the menu bar and choose **Chart.** The Chart Dialog box shown in Figure 4.4 will appear.

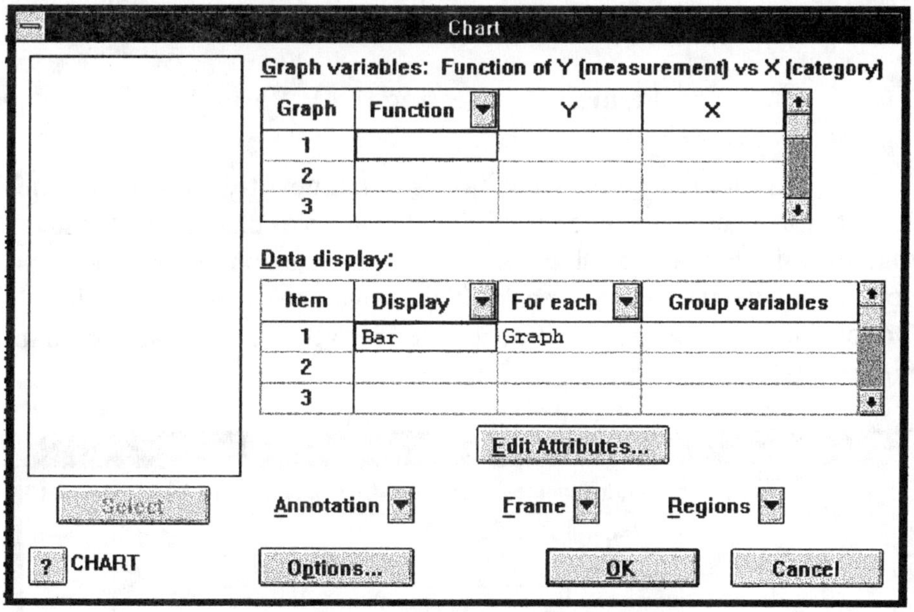

Remember! To select an item from the menu click on the item or use the ⬇ to move to the item and hit ⏎

Figure 4.4 The Chart Dialog Box

The Chart dialog box has two sections that require input. The first is the **Graph Variables** which is where you specify which variables you want to graph and the second is **Data Display** which is where you specify what type of chart you want to use.

Move the cursor to the text box under **Y** and click on the box. In the rectangle at the left of the dialog box (list box) a list of all of the variables that can be used for the Y variable (measurement) will appear. You see that only C2 (Year) and C5 (Number) and C7 and C10 which are copies of those variables are valid as input. That is because the Y variable is a ***quantitative (numerical)*** quantity and the other variables in the data set are ***qualitative (category)*** variables. In this case we want to use C10, which is a copy of *Number* as the Y variable. Double click on C10 from the list box on the left and it will be inserted into the text box under **Y**. Your dialog box should look like the one is Figure 4.5.

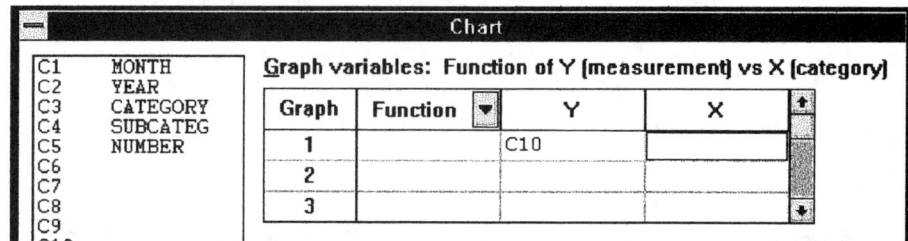

Figure 4.5 Selecting the Y Variable.

The cursor automatically moves to the next text box, under the **X** (category) variable. The list of eligible variables changes to those that are valid for this variable. Since we want a graph of customer complaints we will use the variable *Subcategories* as the X variable. Double click on C9 and it will be inserted into the text box as shown in Figure 4.6.

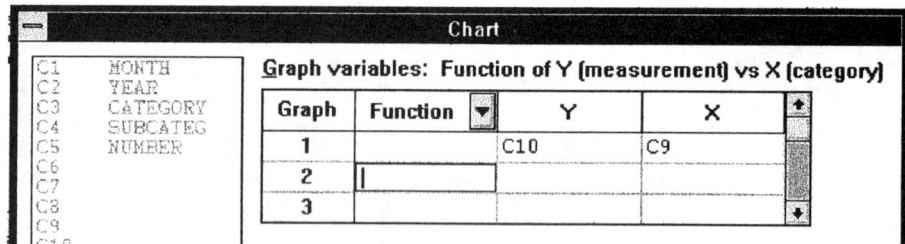

Figure 4.6 Selecting the X Variable

The next thing that Minitab needs to know is what you want done with the data contained in the Y variable column. This may seem strange since you simply want to use the data as it appears, but since Minitab is capable of making bar charts from raw data, that is, a data file in which every line is the data for an individual complaint, it does make sense in a general way. In the dialog box you will see the item **Function** to the left of the Y and X variable columns. Click on the arrow next to the word **Function** to make the Function pop-up menu appear. This menu is shown in Figure 4.7.

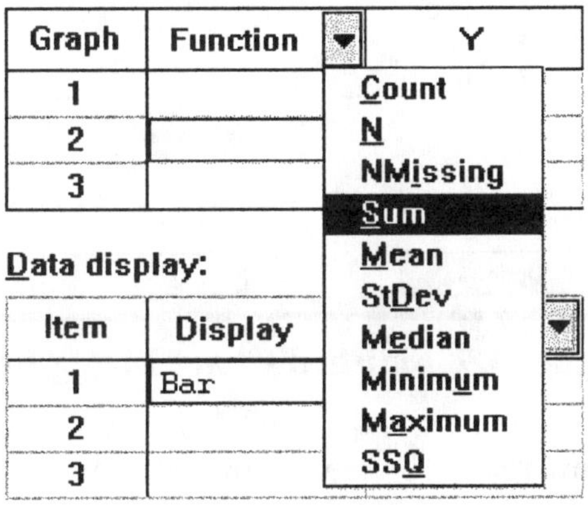

Graph	Function ▼	Y
1		Count
2		N
3		NMissing
		Sum

Data display:

Item	Display	Mean
		StDev
1	Bar	Median
2		Minimum
3		Maximum
		SSQ

Figure 4.7 The Chart Function Menu

From the menu select **Sum**. Since we have only one line for each category, the sum of the numbers for each category will simply be the data value itself.

The last thing you need to do is to indicate what type of graph you want to display. In the table labeled Data display, click on the arrow next to the word **Display** to view the Display pop-up menu shown in Figure 4.8.

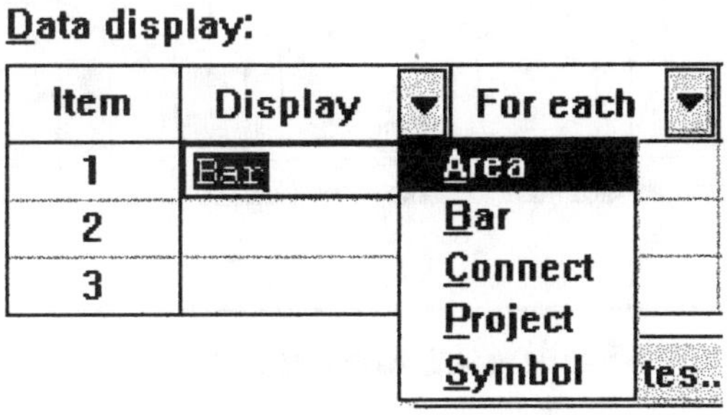

Data display:

Item	Display ▼	For each ▼
1	Bar	Area
2		Bar
3		Connect
		Project
		Symbol

Figure 4.8 The Display Menu

The Chart Dialog Box should look like the one shown in Figure 4.9.

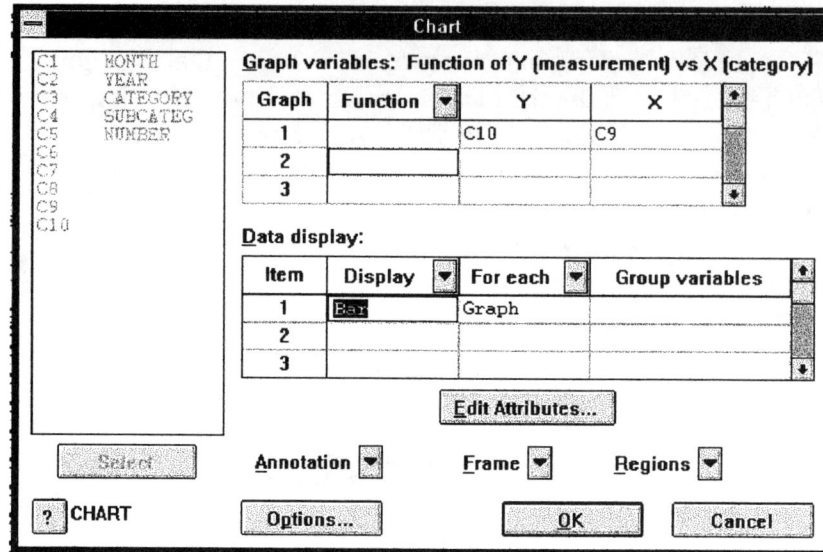

Figure 4.9 Finished Chart Dialog Box

You are now ready to view your first graph! Select **OK** and the graph will display in a window on the screen like the one in Figure 4.10.

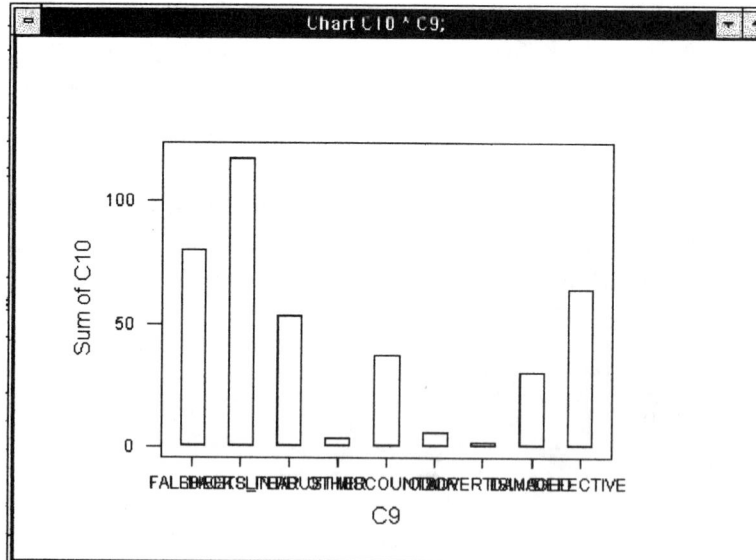

Figure 4.10 First Bar Chart

Exercise 1. What does the bar graph tell you about the distribution of customer complaints? Which category is the highest? the lowest?

Section 4.2.2 Adding Titles to the Chart

Clearly this is not an acceptable finished product! Since the purpose of statistics is to convey **information** about the data we need to clean up the x axis and add titles and other amenities to the graph.

Minitab has two methods for adding titles and editing graphs, through the dialog box for the graph or by selecting **Editor>Edit** from the menu bar that appears when the chart is displayed. For this chapter you will use the dialog box method because that method retains the changes for future graphs.

NOTE: In this book sequential menu commands appear as the commands separated by the ' > ' symbol

Close the current graph window and select **Graph > Chart** from the menu bar again. Notice that when you reopen the Chart dialog box, the input from your previous graph is still there.

*Remember to close a window, click on the Control Menu box and select **Close** or just double click on the Control Menu Box.*

At the bottom of the chart dialog box there are you see three pop-up menus, **Annotation, Frame and Regions.** Each of these pop-up menus present options for configuring your graph. The **Annotation** list, shown in Figure 4.11 controls features of graphs concerning text items and display format.

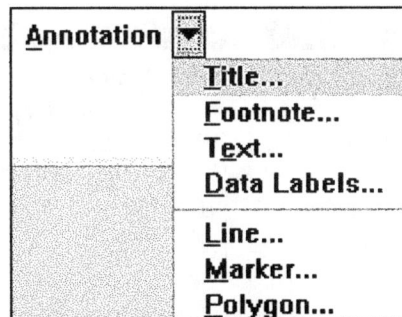

Figure 4.11 The Annotate Menu

To add a title to the graph select **Title** and the Title dialog box shown in Figure 4.12 will appear.

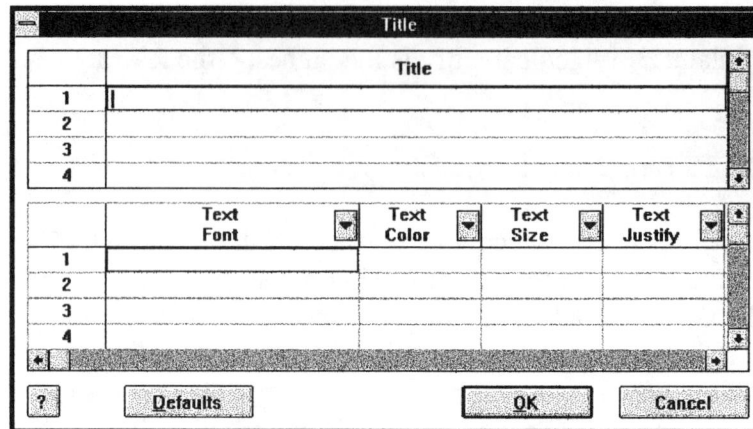

Figure 4.12 The Title Dialog Box

Remember!, In a dialog box ,do not use the [ENTER] key to move to the next line or field. That will select OK and close the box. Use the [↓] and [TAB] keys instead.

Minitab allows you to add four titles to a graph. Position the cursor in the first text box under the word **Title** and type the title you want to appear. If you want a second title, use the text box labeled '2'. The bottom half of the dialog box allows you to specify how each title will appear. For the moment accept the default values. Select **OK** to close the dialog box.

Exercise 2. Choose appropriate titles for the graph you just created and enter them now. Then, exit the **Title** dialog box and redisplay the graph by selecting **OK.** Does it resemble the one shown in Figure 4.13? (It should!)

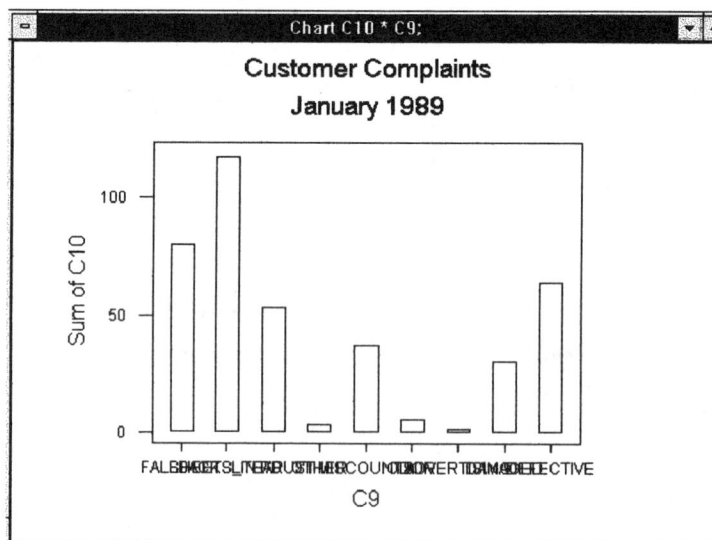

Figure 4.13 Bar Chart with Titles

Section 4.2.1 Modifying a Graph in Minitab

Adding Axis Labels

There are still obviously some problems with the appearance of the graph. The labels on the x and y axes are not very informative and the category labels on the x-axis are jumbled together and unreadable. While this might not *seem* important, it *is* important since the current graph does not provide useful information. In addition to adding titles to the graph, Minitab allows you to control other aspects of the graph's appearance. The **Frame** pop up menu shown in Figure 4.14 controls features of graphs concerning axes such as scaling and tick marks and gridlines.

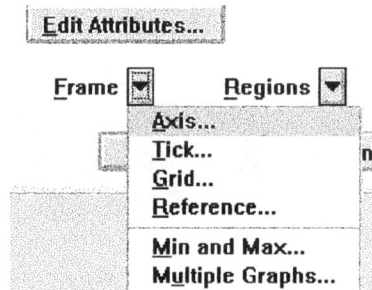

Figure 4.14 The Frame Submenu

Select **Axis** from the **Frame** pop-up menu and the dialog box shown in Figure 4.15 will appear.

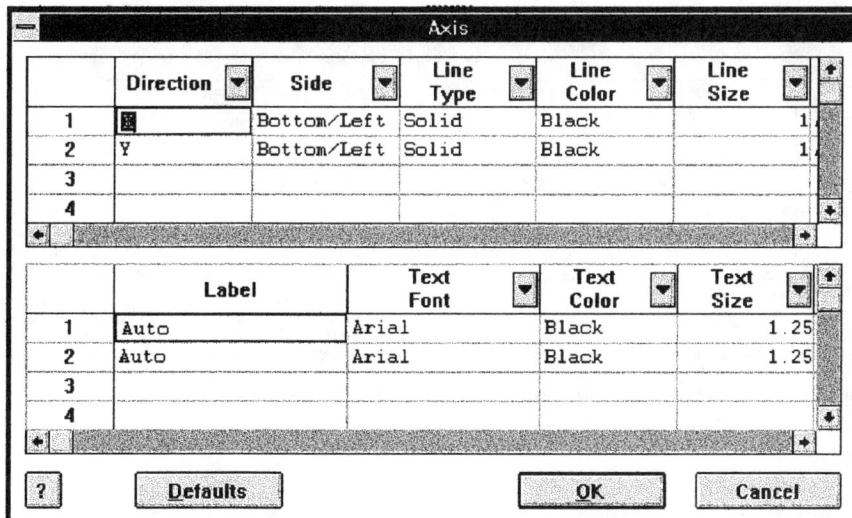

Figure 4.15 The Axis Dialog Box

At the top of the dialog box you see that line 1 is assigned to the X axis and line 2 is assigned to the Y axis. The dialog box gives you the ability to change almost all of the characteristics of the graph axes. For this graph we need to add labels to the axes which indicate what is being displayed on each axis. Position the cursor in the text box under the word **Label** in line 1 and type in a description of the X variable, such as "Complaint Subcategory". Do the same thing for the Y axis. Click on **OK** to return to the Chart Dialog box.

As an alternative to using the Axis Dialog Box, you could name the columns you created. The chart automatically uses the variable names.

If you want to view the graph at any point to make sure that you are satisfied with the changes click on **OK** in the Chart Dialog box. To return to the editing session, close the graph window and select **Graph>Chart** . You will be returned to the current Chart Dialog box with all of your settings intact.

Changing the Tick Marks

The rest of the changes we want to make are associated with the tick marks on the axes. Select **Frame>Tick** and the dialog box shown in Figure 4.16 appears.

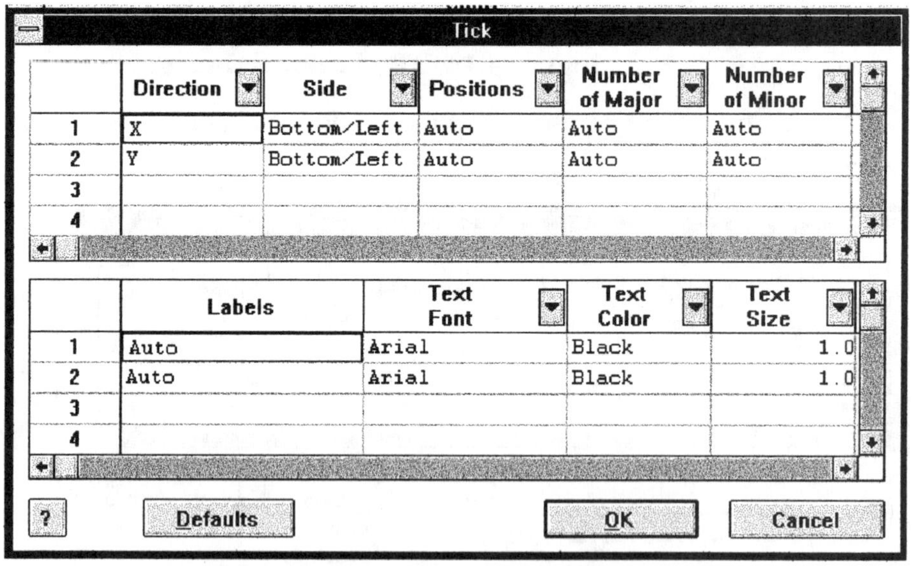

	Direction	Side	Positions	Number of Major	Number of Minor
1	X	Bottom/Left	Auto	Auto	Auto
2	Y	Bottom/Left	Auto	Auto	Auto
3					
4					

	Labels	Text Font	Text Color	Text Size
1	Auto	Arial	Black	1.0
2	Auto	Arial	Black	1.0
3				
4				

Figure 4.16 The Tick Dialog Box

Again you see at the top of the dialog box that Line 1 is associated with the X axis and Line 2 with the Y axis. The problems with the graph are associated with the X axis. The *Subcategories* are unreadable because the text size is too large and they run into each other. We will fix this problem by rotating the tick mark labels and reducing the font size.

At the bottom of the **Tick** dialog box there are several items that can be changed for each axis. The fourth column is used to adjust **Text Size.** Position the cursor in Line 1 under Text size. The current size is 1.5. If you click on the arrow button next to text size, the pop-up menu shown in Figure 4. 17 appears.

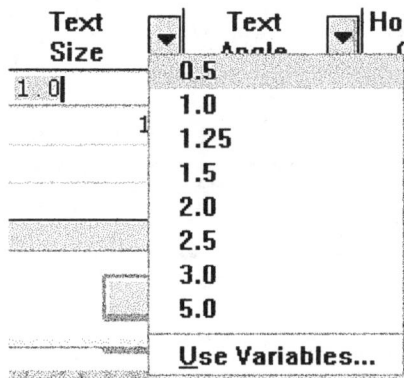

Figure 4.17 The Text Size Pop-up Menu

From this menu select 0.5 (the smallest size). You are not limited to the selections offered on the menu, but they are usually adequate. Now, use the ▣ key (right arrow) to move to the right to the column labeled **Text Angle**. This option allows you to rotate the text from the horizontal (0). Click on the arrow to display the pop-up menu and select **90** to rotate the labels 90 degrees.

When the text is rotated, it is done around the center of each label. This usually results in the labels being placed too high on the graph. To fix this problem you need to adjust the *offset* of the labels. Use the ▣ to move two text boxes to the right, to the column labeled **Vertical Offset**. This is set to **Auto** which is the default. You want to move the labels down, which is the negative direction. Click on the arrow and from the menu select -0.05 which is the most that you can move the label in that direction. If you set one offset manually you cannot use **Auto** for the other. Use the ▣ to move to the **Horizontal**

Offset column and from the pop-up menu select 0. This will not change the horizontal position, which is what we wanted.

When you have made these changes the **Tick** dialog box should look like the one in Figure 4.18.

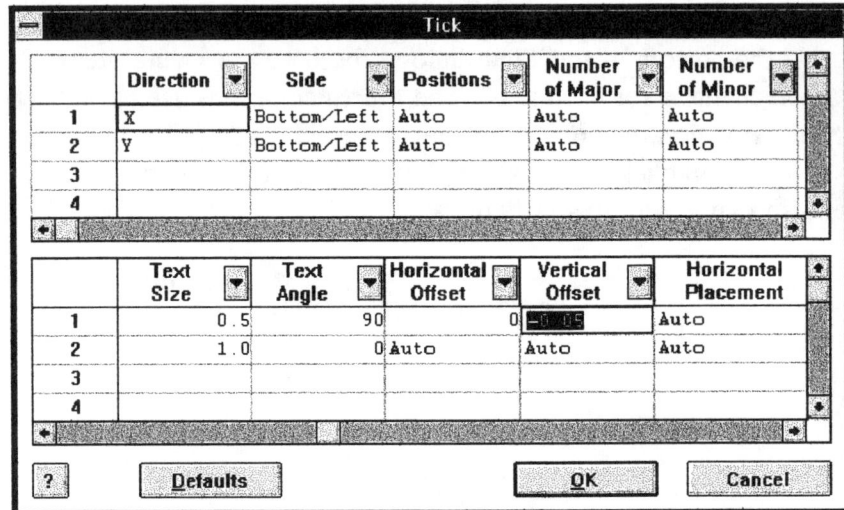

	Direction	Side	Positions	Number of Major	Number of Minor
1	X	Bottom/Left	Auto	Auto	Auto
2	Y	Bottom/Left	Auto	Auto	Auto
3					
4					

	Text Size	Text Angle	Horizontal Offset	Vertical Offset	Horizontal Placement
1	0.5	90	0	-0.05	Auto
2	1.0	0	Auto	Auto	Auto
3					
4					

Figure 4.18 Finished Tick Dialog Box

Now select **OK** to return to the **Chart** dialog box and **OK** to display the chart. It should look like the one in Figure 4.19.

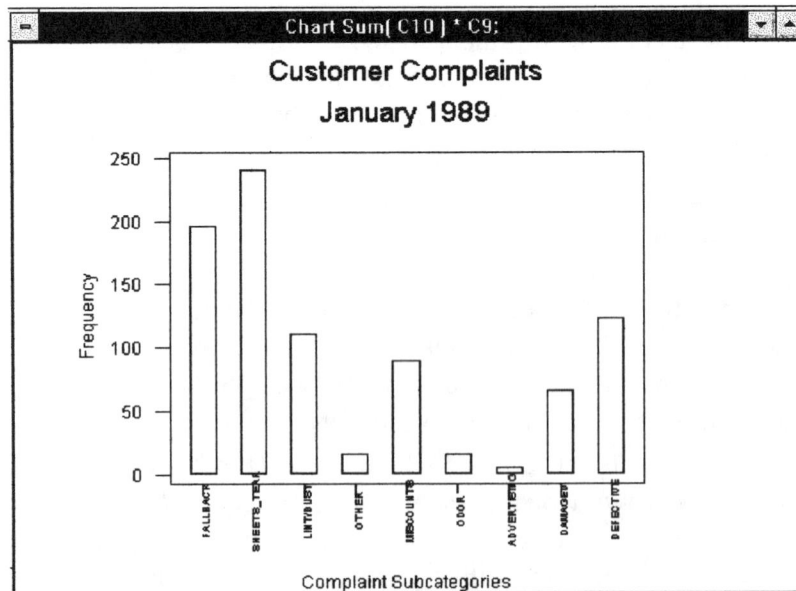

Figure 4.19 Finished Bar Chart

It may seem that we have spent a good deal of effort to make this chart, but the commands you learned here are the same for almost all of the different kinds of graphs you can create in Minitab.

Another, more basic problem is that the bar chart created has too many categories to be a good bar chart. The number of categories used should not exceed 7 or 8 in order for the graph to be effective, and this one has 10. As an exercise later on you will create a Bar Chart using the *Category* variable which will address this problem.

Section 4.2.3 Saving, Opening and Printing Graphs

Saving Graphs

Before we move on to different methods of displaying our complaint data, you may want to save your graph so that you can view it or print it out again at another time. To do this you should *save* the graph.

To save a graph in Minitab, select **File>Save Window As** and the **Save Graph As** dialog box shown in Figure 20 appears.

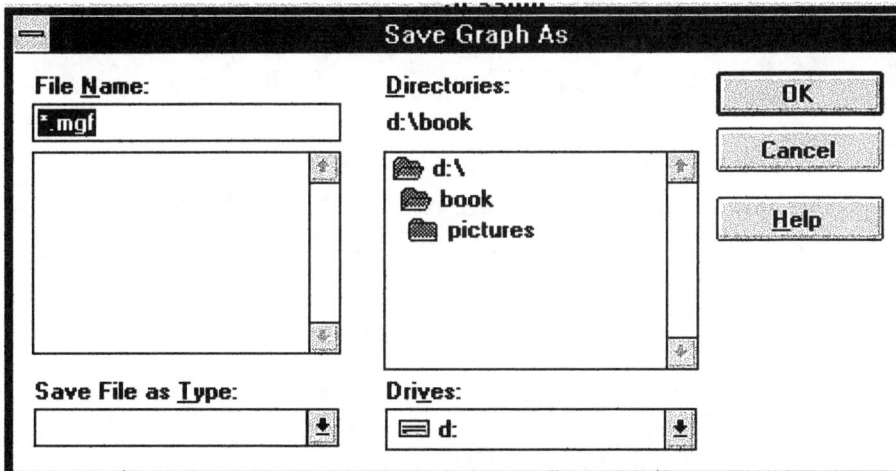

Figure 4.20 "Save Graph As" Dialog Box

In the text box labeled **File** Name type the name of the file for the graph. Use something that will help you identify the graph at a later date, such as jan89. The file suffix for a Minitab graph is .mgf and it is the default value automatically added. In the text box labeled

The x axis labels are still a bit high. Remember that you are not restricted to the offset values in the menu. You can change the value from -0.05 if you want to move the labels

*You may also want to select a particular directory on that drive. Use the **Directories** box to do this. If you are not sure how to do this, consult your Windows documentation.*

Drives: press on the arrow and the drop down list of available drives will appear. Select the drive you want to save the graph to. Click on the command button labeled **OK** and the file will be saved to that drive.

Opening Graphs

If you want to work with a graph that you have already saved select **File** > **Open Graph** and the dialog box shown in Figure 4.21 appears.

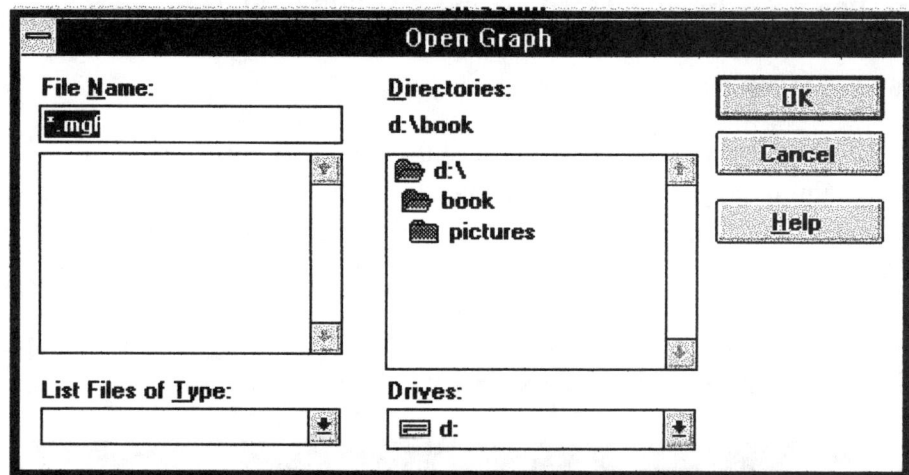

Figure 4.21 The Open Graph Dialog Box

Use the **Drives:** drop-down list to select the drive where the graph is located and the **File Name:** text box to select the file you want. Click on **OK** to open the selected graph.

Printing a Graph

In Minitab you can either print the current graph or any other open graph. To print the current graph select **File** > **Print Window**. The dialog box shown in Figure 4.22 appears. Click on **OK** and the graph will print to the selected printer.

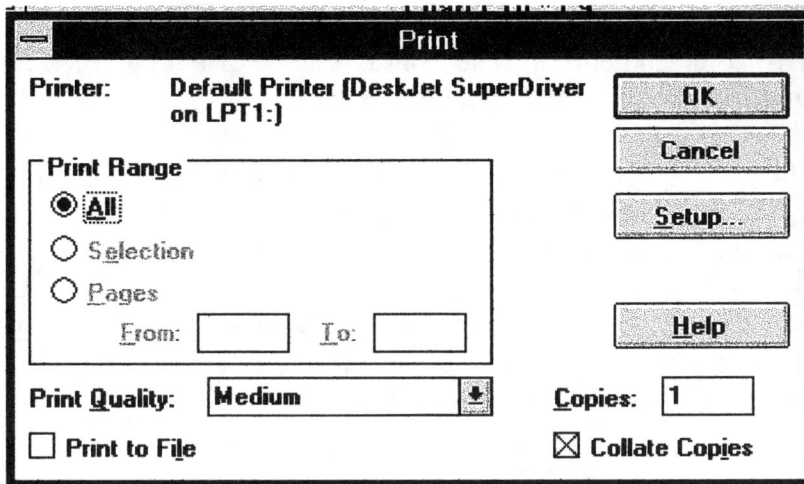

Figure 4.22 The Print Dialog Box

If you have several graphs open and the one you want to print is not open, select **Window** > **Manage Graphs**. The dialog box shown in Figure 4.23 will appear.

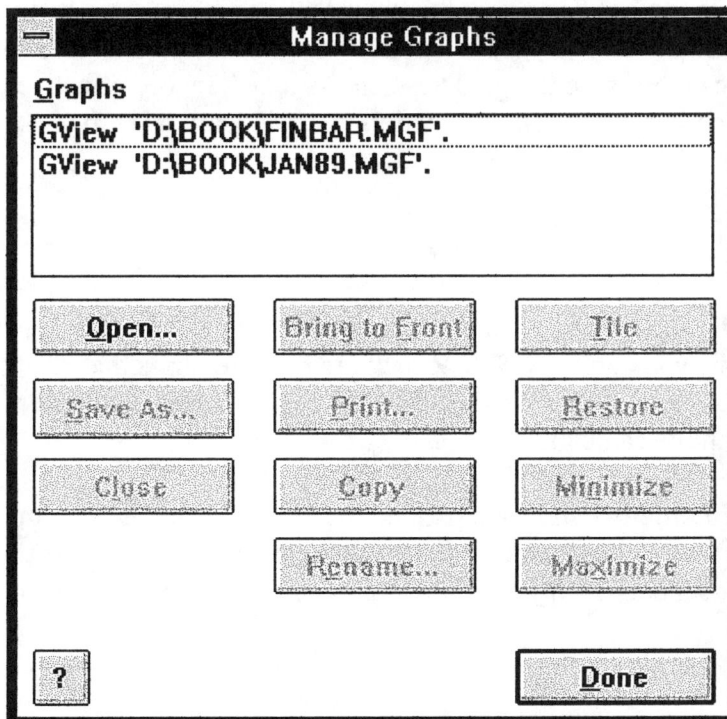

Figure 23 The Manage Graph Dialog Box

*NOTE: You will be
using the data for
January 1989 many
times in this chapter.
It is probably a good
idea to leave this data
in one place and use a
different set of empty
columns for the
Exercises.*

Use the cursor to select the graph you want to print and click
on the **Print** command button. The Print Dialog box from Figure 4.22
appears and you can save the graph the same way.

The **Manage Graphs** dialog box allows you to perform a
number of different actions with your graphs. It will be especially
useful as the number of graphs you create increases.

Exercise 3   Create bar charts for the defect sub categories for January
1990 and January 1991. Be sure to include the titles and save the
graphs. You will need to refer to them later on.

Section 4.6 Clustered and Stacked Bar Charts

Comparing data for several different scenarios is something
that management needs to do in order to make sound business
decisions. If, for example, the company wanted to know whether
complaints for the last month were comparable to previous months,
they could look at bar charts for the months of interest together to get
this information. It is possible to use other types of bar charts to
accomplish the same thing. These charts may be more effective for
making comparisons. In this section you will learn how to view your
data using CLUSTERED and STACKED bar charts.

Section 4.6.1 Clustered Bar Charts

In a bar chart, when you have similar data for different
situations, you can **CLUSTER** the bars for categories that are the
same together, so that the data from all of the different situations
appear on the same graph.

In order to make a clustered bar chart for the fist two months
of 1989 you will use most of the same steps that you used for the bar
chart for January. Copy the data from February 1989 to the same

columns that you copied January 1989. Locate February below
January. If you deleted the copy of January 1989 copy both January
and February 1989 to a new set of columns.

Select **Graph** > **Chart** to open the Chart dialog box. Using
the same steps you used for the first bar graph, fill in the variables that
will be used for the X and Y variables and select **Sum** as the Function.

The section marked **Data Display:** is used to tell Minitab how
you want the graph to appear. Since you want a different bar for each
month, from the **Display** column select **Bar** and from **For Each:**
select **Group**. The next column, **Group Variables** indicates the
variable you want to use to define the groups. When you locate the
cursor in the column the list of valid grouping variables is displayed in
the list box at the left. Select **C6**, which contains the variable *Month*.
Your dialog box should look like the one in Figure 4.24.

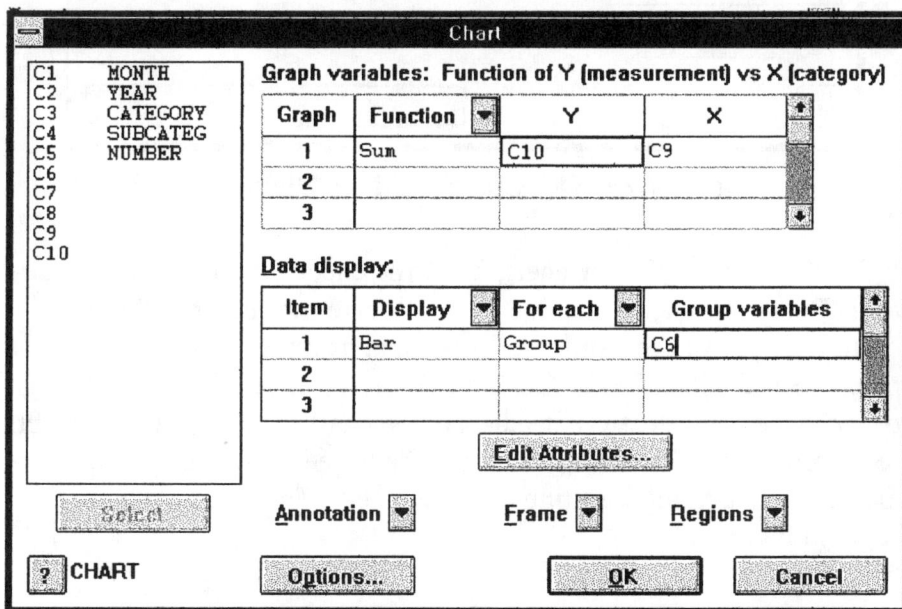

Figure 4.24 Chart Dialog Box for Clustered Bar Chart

There are several options for displaying bar charts with grouped variables. Select the **Options...** command button and the dialog box shown in Figure 4.25 appears.

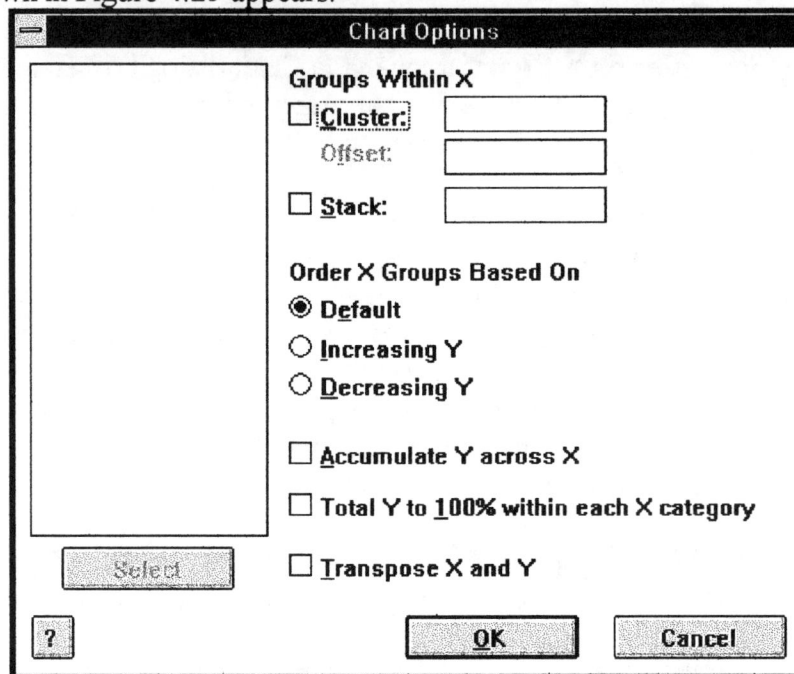

Figure 4.25 The Chart Options Dialog Box

To create a **cluster** chart, mark the check box to the left of the word **Cluster** by clicking on it and then locate the cursor in the text box to the right of the word. A list of valid group variables will appear. Select C6, *Month*. Click on **OK** to return to the Chart dialog box. *Before* you select **OK** to display the chart, change the titles to fit the current data. The settings that you have been using for the axis labels and the tick mark settings are still valid. The finished chart is shown in Figure 4.26.

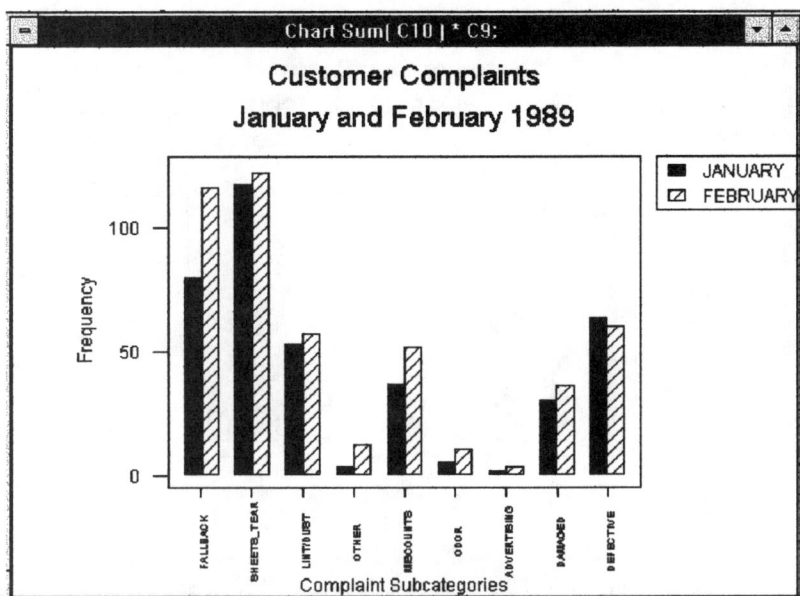

Figure 4.26 Clustered Bar Chart for January and February 1989

Exercise 4. Create clustered bar charts for January and February of 1990. Repeat the exercise for 1991.

Section 4.3.2 Stacked Bar Charts

Rather than having the bars of the chart clustered so that you can do direct comparisons among months or look at distributions by months, you may be interested in another way of looking at the information you have. A Stacked Bar Chart can be used to see the way each month contributes to the overall total number of complaints in a particular category. To create a Stacked Bar Chart with the same data you just used for the Clustered Bar Chart, you only need to make one change. Select **Graph** > **Chart** and settings for the last clustered bar chart you make will appear in the Chart dialog box. Click on the **Options...** command button and change the **Groups Within X** setting from **Clustered** to **Stacked**. Click on **OK** to return to the Chart dialog box and again to display the graph. It should look like the one in Figure 4.27.

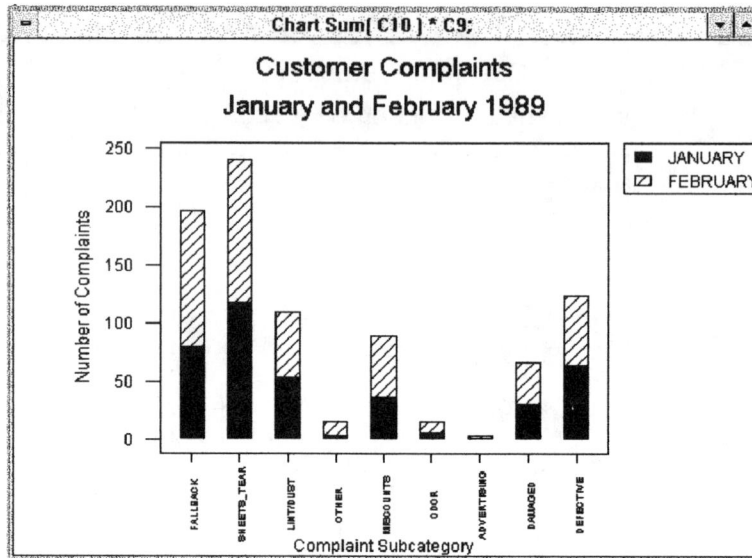

Figure 4.27 Stacked Bar Chart for January and February 1989

Section 4.7 How to Create a Pie Chart in Minitab

Qualitative data can also be displayed using a Pie Chart. A pie chart is particularly useful when you want to look at different categories as a percent of some total. To make a pie chart for the January 1989 data, copy that data to a set of empty columns in the worksheet. Select **Graph** > **Pie** and the Pie Chart dialog box shown in Figure 4.28 appears.

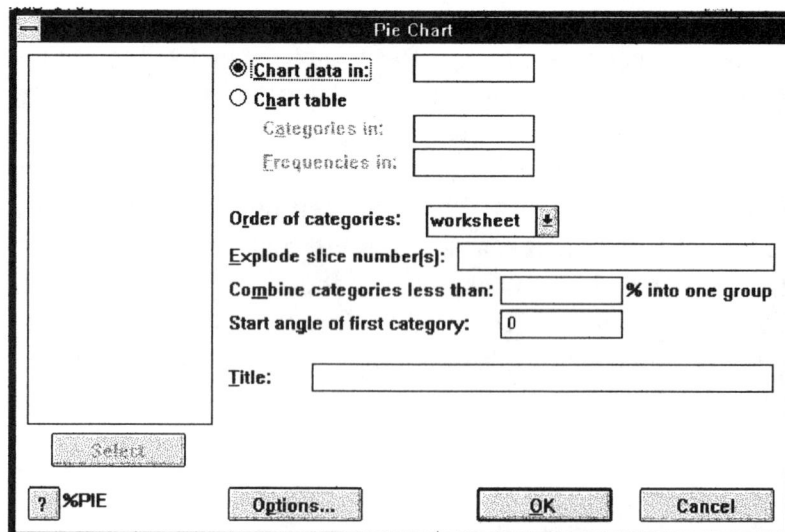

Figure 4.28 The Pie Chart Dialog Box

As with bar charts, Minitab allows you to create pie charts from raw data or from a frequency table. To create a pie chart from a table, click on the options button for **Chart table**. Position the cursor in the text box labeled **Categories in:** and from the list box, select C9. Move the cursor down to the text box **Frequencies in:** and select C10. Move the cursor to the text box labeled **Title** and type in a title for the chart. The completed dialog box should look something like the one in Figure 4.29.

Note! Minitab will not allow categories to have relative frequencies less than 0.1%. If you have categories with 0 or very small frequencies you will need to combine these categories!

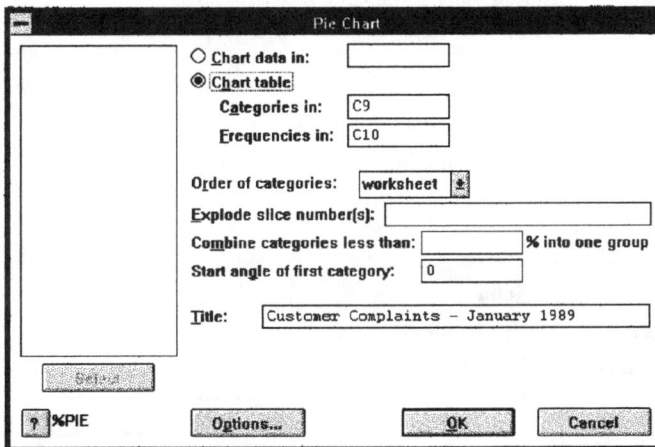

Figure 4.29 Completed Pie Chart Dialog Box

For the moment, accept all of the other default values for the chart and click on **OK**. The chart shown in Figure 4.30 should appear.

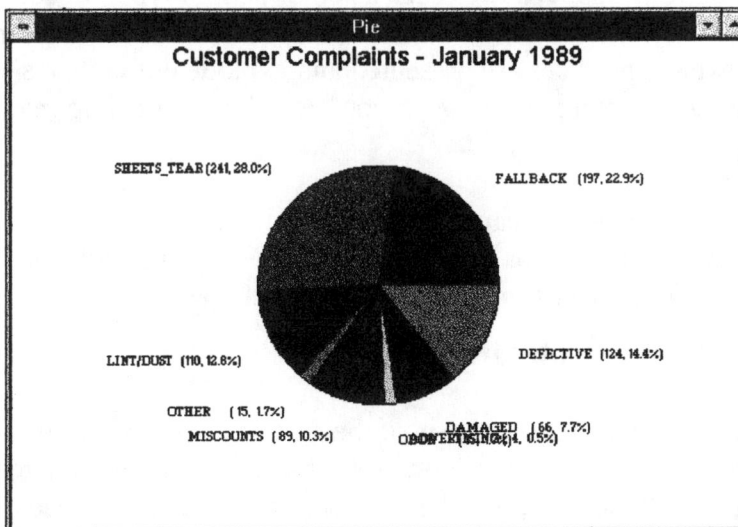

Figure 4.30 Pie Chart for January 1989 Complaint Data

There are several parameters that you can change to alter the appearance of the chart. Close the graph and select **Graph** > **Pie** and click on the **Options...** command button. The Pie Chart Options dialog box shown in Figure 4.31 appears.

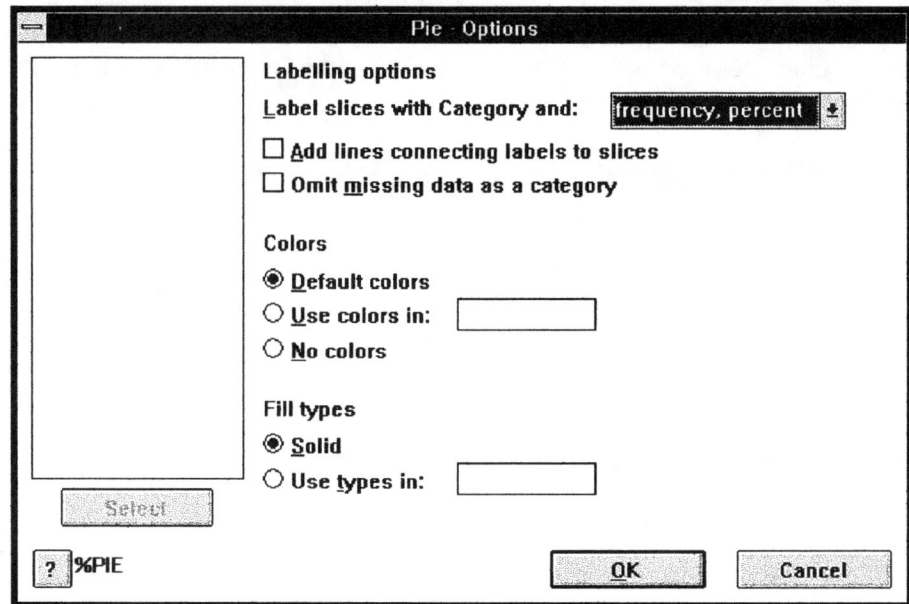

Figure 4.31 Pie Chart Options Dialog Box

This dialog box allows you to change the way the pie slices are labeled and to change features such as color and fill type. In addition to these changes, from the Pie Chart dialog box you can change the order in which the slices are presented and explode out certain slices. These features are important when you are using to graph to make a certain point.

Exercise 5. Create pie charts for the complaint sub category data for January 1990 and then for January 1991. Which do you think is more informative for this data, a pie chart or a bar chart?

Section 4.8 Consolidating Data

Remember that the x-axis of the first bar chart you created was too crowded to read and there was too much information to digest. Although you "fixed" the problems by changing the appearance of the graph, the real problem is that the bar chart had too many classifications to be effective. An alternative is to create a bar chart using the complaint categories instead of the sub categories.

In Minitab this is done by changing the variable for the x axis from *Subcategory* to *Category* and by using Sum as the function. This will tell Minitab to sum the frequencies for all categories that are the same.

Using the data for January 1989 select **Graph** > **Bar**. In the column marked **Function** select **Sum**. Enter **C10** in the Y variable text box and in the X variable text box enter **C8,** which is the *Category* variable. Enter the appropriate titles and click on **OK**. Your chart should resemble the one shown in Figure 4.32.

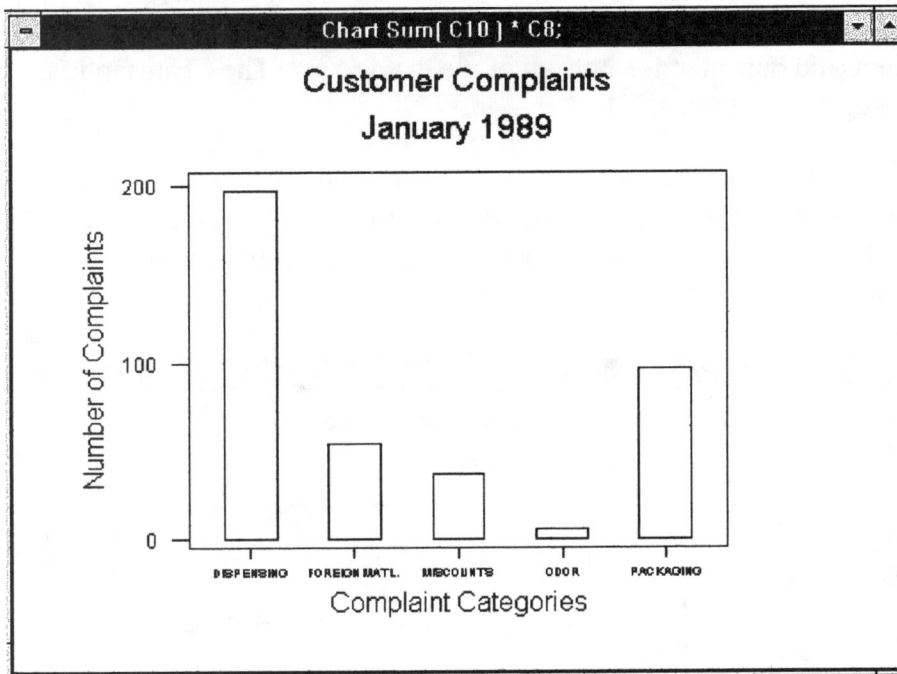

Figure 4.32 Bar Chart for Major Category Defects

It is possible to create bar charts for major defect categories for any subset of the data by copying that data to a new set of columns and using the **Sum** function in the Chart dialog box.

Exercise 6. Create bar charts for the major category classifications for January 1990 and January 1991.

Section 4.9 Pareto Diagrams

The bar chart just created for the major defect categories is an improvement over the previous attempts, but it still leaves something to be desired. Remember that management is trying to identify the primary cause of customer complaints and the while the information is available in our bar chart, it is not emphasized. Data of this type is better displayed with the bars arranged in descending order of frequency. We can accomplish this in Minitab in two ways.

The first method for displaying the bars in the bar chart in a specified order is to use the **Options** command in the Chart dialog box. Select **Graph** > **Chart** and fill in the main Chart dialog box to recreate the bar chart shown in Figure 4.32. When you have done this, **before** you click **OK** to display the graph, click on the **Options** command button at the bottom of the dialog box. The Chart Options dialog box will open.

This is the same dialog box that you used to create clustered and stacked bar charts. It is shown in Figure 4.25.

The center section of the dialog box has a set of option buttons and is headed **Order X Groups Based On.** Select **Decreasing Y** and click on **OK** to close the dialog box. Click **OK** again to display the graph. It should look like the one in Figure 4.33.

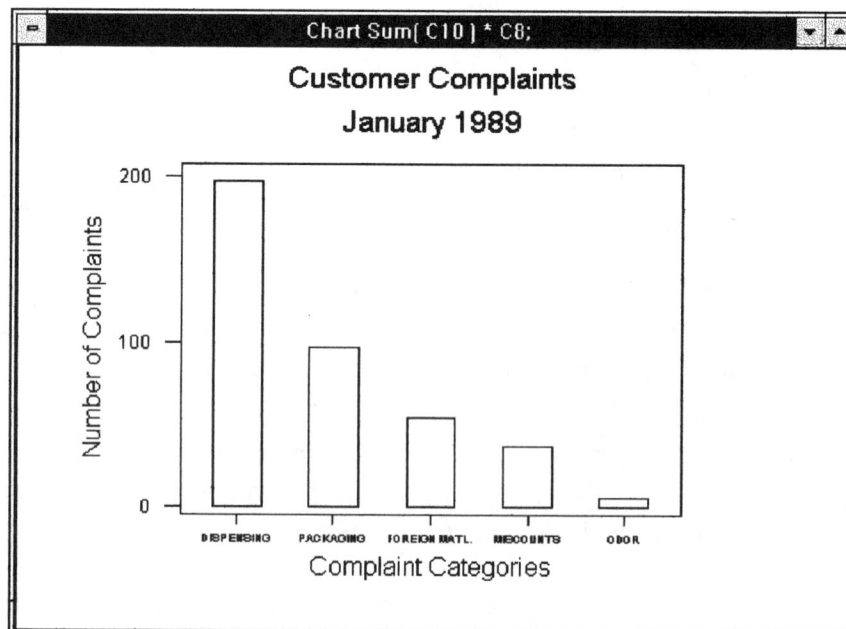

Figure 4.33 Complaint Categories in Descending Order

This bar chart gives management more insight into where efforts to reduce customer complaints should be directed. If the data are to be incorporated into a presentation of any type, this chart will be the most effective in conveying the information.

What you have just created is closely related to a tool from **Statistical Process Control** known as a *Pareto Diagram*. A Pareto Diagram is often used to identify trouble spots in defect and complaint data. The Pareto Diagram is a bar chart with the bars in order of decreasing frequency, but it adds a line that plots the *cumulative frequency* of the categories. This not only tells management which categories are the largest, but gives information about what percent of the total problems are accounted for by the largest categories.

Minitab has a separate option for creating Pareto Diagrams. From the main menu bar select **Stat**. The **Stat** menu is shown in Figure 4.34.

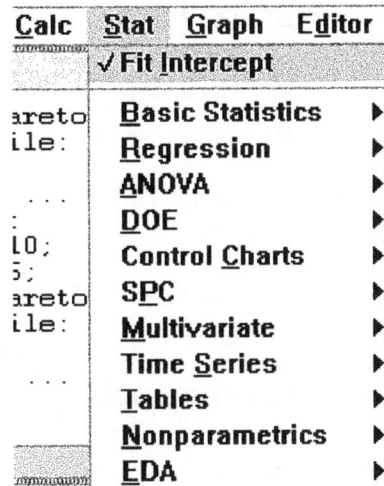

Figure 4.34 The Stat Menu

You can also open a cascading menu by clicking on the menu selection.

Pareto charts are part of the **SPC** (Statistical Process Control) selection. Highlight **SPC** use the ▣ key or hit ⏎ to open the cascading menu shown in Figure 4.35.

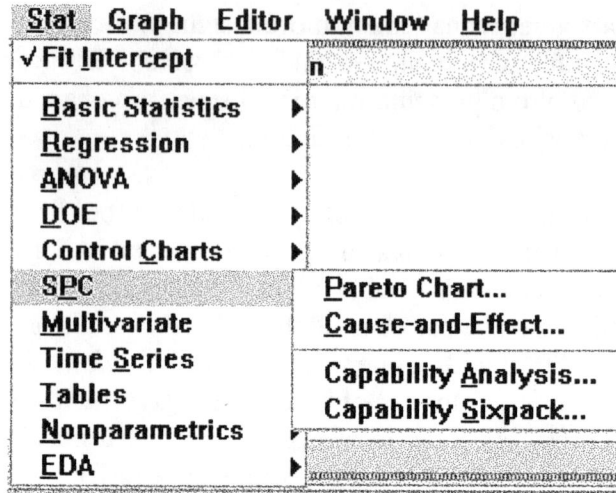

Figure 4.35 SPC Cascading Menu

Select **Pareto Chart** and the dialog box shown in Figure 4.36 will open.

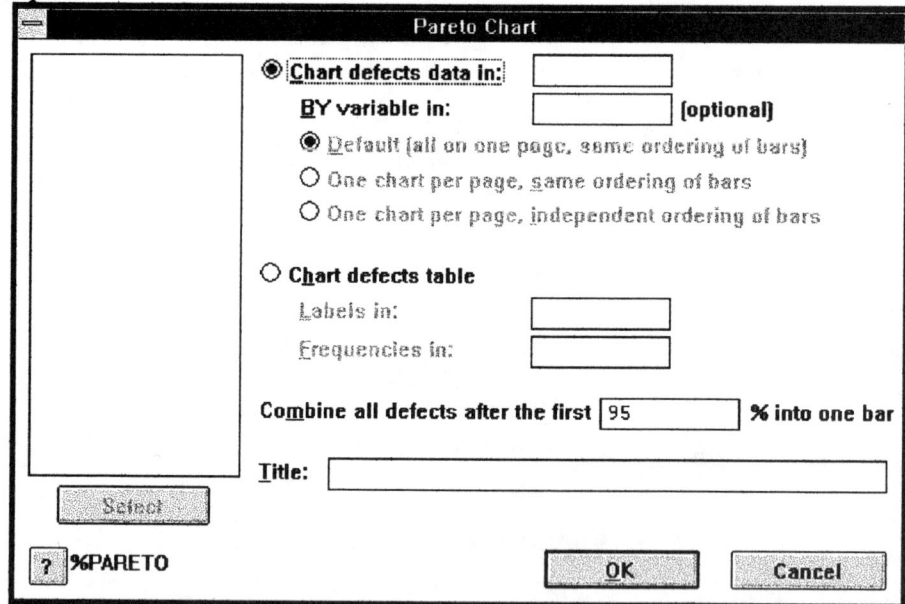

Figure 4.36 Pareto Chart Dialog Box

The **Pareto Chart** feature is like the one that creates pie charts in that it will accept the data in raw or tabulated form. Mark the option button labeled **Chart defects table** by clicking on it and

position the cursor in the text box to the right of the words **Labels in:**.
From the variables selection list choose **C8**. Next select **C10** for
Frequencies in:.   Type in an appropriate title and click on **OK** to
display the chart shown in Figure 4.37.

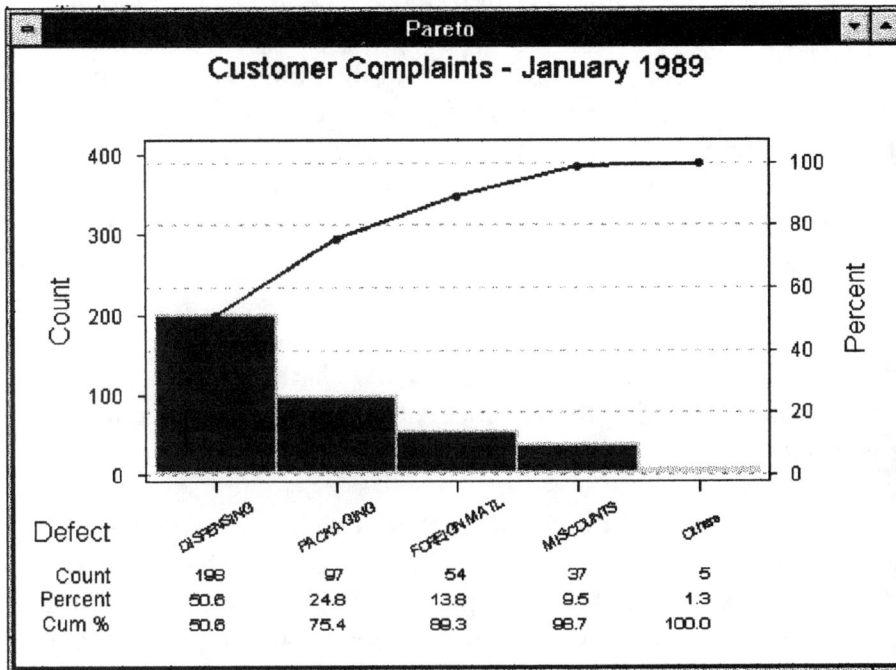

Figure 4.37 Pareto Chart for January 1989

The left scale of the Pareto Chart gives the frequencies for each
individual category while the right y scale gives the cumulative percent
of the total for a category and all of the categories to the left of it.
From Figure 4.37 you see that Dispensing accounts for 50.6% of the
total complaints for January 1989, while Dispensing and Packaging
together account for 75.4%. This information is useful in targeting
process problems to reduce complaints. If the company can determine
how to correct the problems that cause the dispensing complaints they
can cut the number of complaints in half! Even if they cannot
eliminate all of the problems this category provides the most *potential*
for reducing complaints.

Exercise 7. Create a Pareto chart for January 1990 and compare it to
the one for January 1989? How are they the same? How are they
different?

Section 4.10 Looking at the Data for Different Time Periods

So far you have looked at the data in the time frame that it was collected, a month at a time. With so much data available, it would not make sense to make decisions based on any single month. In fact there is no real reason to believe that all of the months are similar. When data is available for many different time periods it is usually intelligent to look at the data for these different periods in several different ways. For example, the company might be interested in looking at the trend of complaints over an entire year to see if there is any information to be obtained.

Since the bar chart option allows you to sum the data over the categories, all that is required to create a bar chart that looks at monthly data is changing the X variable (the classification) from *Category* to *Month*.

Copy the rest of the data for 1989 to the rows below the data for January 1989. Select **Graph** > **Chart** and set up a bar chart using **C6** for the X variable and **C10** for the for the X variable. Be sure to select **Sum** for the Function. Fill in the appropriate titles and display the graph. It should look something like the one in Figure 4.38.

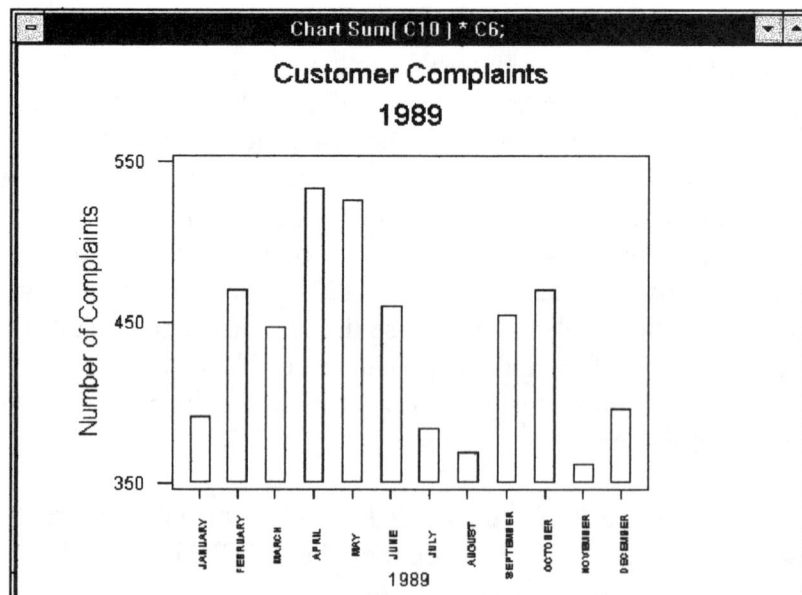

Figure 4.38 Complaint Data for 1989

Exercise 8. Do you notice any interesting features about the number of complaints over the year? Can you offer any explanations for why this might occur?

Exercise 9. Create a bar chart that shows the total defects by month over the three year period. Compare this graph to the one in Figure 4.38. Are the two years similar? If not, how are they different?

Section 4.11 Investigative Exercises

In the following exercises you are asked to use the skills introduced in the previous sections to extract information from the complaint data file. You are provided with space to answer the questions and paste in graphical output from the program. If you do not have access to a printer, you can sketch the graphs on the axes provided.

1. a) Construct a bar chart of the packaging sub categories complaints for January, 1990.

b) Construct a pie chart for the same data.

c) What packaging sub category contributes the most complaints? Which is second? Which graph shows this better? Why?

2. a) Create a bar chart for packaging sub categories for January 1991.

b) Do the same thing for October 1990.

c) Do the distribution of defects appear to be the same for the two graphs that display January data? Would you expect them to be?

d) Is the distribution for October similar to these two? Would you expect it to be?

3. Create pie charts for the same three months.

b) Is it easier to make comparisons using bar charts or pie charts?

4. a) Create a clustered bar chart for the same three months.

b) Create a stacked bar chart for the same data.

c) Are comparisons like this easier using a single graph or three different graphs? What things do you have to be careful of when you use three different graphs?

5. a) Calculate the total number of packaging complaints for each month of 1989.

b) Create a bar chart that displays this data.

c) Examine the bar chart you created in part b. What do you notice about the number of packaging complaints over the year? Is it constant? If not, what do you see happening?

d) Create a pie chart for this data? Can you answer part c using the pie chart? Why or why not?

6. a) Repeat exercise 4, parts a and b for the packaging data from 1990 and then for 1991.

b) Make a clustered bar chart of the packaging data from the three years.

c) Compare the behavior of the number of packaging complaints over the years. What do you observe?

7. a) Look at the total number of complaints by month for 1990 and 1991.

b) Make bar charts of the monthly totals for these two years.

c) Compare these graphs to each other and to the graph for 1989 that you created earlier in the chapter.   Are there any noticeable trends or patterns? What differences or similarities do you see among the years?

d) What months or seasons generally have the most complaints? The fewest?

e) Can you offer an explanation for this?

8. Compare the monthly bar charts for packaging defects to the ones for total defects. Are they similar ? Would you expect them to be related to each other? Why or why not?

9. a) Create pie charts for total complaints using the variable *Category* for each of the years.

b) Have there been any major shifts in the relative frequency of the types of complaints over the past three years?

c) Display the same information on a clustered bar chart and a stacked bar chart.

d) Which graph do you think is more informative? Why?

e) Do you get the same information from these display that you get from the pie charts? Which type do you think will give the company a better picture of their complaint history? Why?

10. a) Suppose you wanted to compare the distribution of complaints (by category) on a monthly basis to see if there are any similarities. What type of graph(s) would you use? Why?

b) Do this for one of the months. What is your conclusion?

11. The management of this company asks you for your assessment of their complaint data. In particular they would like to know where their biggest problems are and what the trends seem to be over the past three years. Create any additional graphs that will help you answer their questions and prepare a report for them offering your analysis and your suggestions.

Chapter 5   "Golf Ball Design"
Displaying Quantitative Data

Section 5.1 Overview

Statistical Objectives: After reading this chapter and doing the exercises you will:

- Know the function of a histogram.
- Know how to describe the shape of a distribution from the histogram.
- Know the effects of changing class interval widths.
- Know the effects of changing the number of class intervals.
- Know how to compare variables using histograms.
- Know how to detect outliers using histograms.
- Know how to detect trends in data using histograms.

Section 5.2 Problem Statement

All companies are currently facing increasing competition at the national and international level. In the face of this competition American manufacturers are moving to a focus on *quality*. This is not a simple change but rather encompasses all aspects of the business and all employees of the company. Managers are learning to listen to the creative ideas of their employees, are breaking down the hierarchical management layers and monitoring their production processes in order to prevent problems ahead of time. In doing so, more and more data is being collected on every aspect of the business. This includes data on such things as customer complaints as well as data on how closely the product matches with the target design. It means all parts of the organization must work together as a team to achieve Total Quality Management (TQM).

In this chapter you will look at how a large manufacturer of golf balls is incorporating some of these issues. Many of you enjoy taking to the golf course at the first hint of good weather. In some parts of the country you can golf every day of the year while others must patiently wait for the snow to melt. Next time you hit the golf ball, think about what type of decisions went into the production of that golf ball in order to make you look good on the golf course -- after all we all want our balls to fly a great distance when we hit them! What can the manufacturer do which ultimately will affect how far the ball flies?

In this particular case the company is trying to evaluate two different ball designs. Which of them will repeatedly perform the best on the golf course?   Anyone who has hit a golf ball knows that in addition to the particular ball design, there are many factors that could influence how far the ball flies; for example the expertise level of the golfer, the wind speed, etc.

In fact, there are many factors that effect how far the ball flies and in order to remain competitive, most golf ball manufacturers are constantly creating and testing new ball designs. Such things as ball size, weight, dimple pattern are all within the control of the manufacturer and are called **internal factors**.

Figure 5.1 A Typical Golf Ball

Such things as the expertise level of the golfer, head speed, launch angle, wind speed, wind direction, relative humidity and temperature also effect the distance the ball flies but are not typically within the control of the manufacturer! These are called **external factors.**

The company is primarily interested in measuring how far the ball carries and the total distance the ball travels. It should be noted that carry is how far the ball has gone from the point where it was hit to where it lands. Total distance also includes the distance that the ball has rolled after it hits the ground. Typically, total distance is greater than carry. Think about how these characteristics can be measured.

This company has a testing site in Florida and balls are placed into machines that hit the balls at a pre-specified speed and launch angle. Many balls are hit by the machine within a short period of time so that factors such as wind speed, wind direction, relative humidity, and temperature are not fluctuating very much. Measurements are taken on these factors to be sure that they are basically constant throughout the test. Thus the external factors are being controlled.

For a given test, the balls used are all the same model number with the same dimple design and they should be the same weight and size (although these are also measured). Thus, the internal factors are being controlled.   The company has attempted to create a situation where all the factors that we listed above, both internal and external factors, are held constant in order to study the characteristics of interest: *carry and total distance*.

NOW the big problem is to measure how far the ball carries! The total distance is fairly easy to measure as we have the ball in its final resting spot as a marker. In order to measure carry we need to know where the ball lands -- not where it finally ends up after rolling a bit. SO -- the company hires people to stand in the fields and place markers where the balls actually hit the ground.   This is true! If you think you have a headache reading this, think about how those field researchers feel!

The data you will be analyzing in this chapter contains information on most of the factors that have just been discussed. Two different ball designs are being compared and the data were taken during three different time periods.

Section 5.3 Characteristics of the Data Set

FILENAME: CHAP5.MTW
SIZE: COLUMNS   14
               ROWS   72

The first 8 columns (C1-C8) of the actual datafile are shown in Figure 5.2.

	C1	C2-A	C3	C4	C5	C6	C7	C8
↓	Ball_	Model _	S1	S2	S3	Wgt	Dw	Dd
1	1	M1	81	81	82	45.3	0.1450	0.0110
2	2	M1	83	83	84	45.2	0.1510	0.0111
3	3	M1	81	82	84	45.2	0.1450	0.0105
4	4	M1	81	81	83	45.3	0.1440	0.0117
5	5	M1	83	81	82	45.5	0.1460	0.0108
6	6	M1	83	83	82	45.3	0.1560	0.0111
7	7	M1	81	81	82	45.2	0.1495	0.0111

Figure 5.2 First 8 Columns of CHAP5.MTW

The remaining 6 columns are shown in Figure 5.3

	C9	C10	C11	C12	C13-A	C14-A
↓	Head	Temp	Carry	Tot Dist	Date	Time
1	686	77	257	270	8/20	8:15
2	688	77	255	267	8/20	8:15
3	687	77	256	267	8/20	8:15
4	688	77	255	271	8/20	8:15
5	687	77	255	268	8/20	8:15
6	687	77	256	267	8/20	8:15
7	687	77	255	264	8/20	8:15

Figure 5.3 Last 6 Columns of CHAP5.MTW

Notes on the data set:

1. The variable *Ball#* keeps track of the observation number and goes from 1 to 72.

2. The variable *Model_* indicates which of the two ball designs is used for that observation. There are two different Model #'s in this dataset:

M1: Ball# 1-12, 25-36, 49-60
M2: Ball# 13-24, 37-48, 61-72

3. The variables *S1, S2,* and *S3* are size measurements. They indicate the measurement of the ball around the ball's equator, and at two other points. All three measurements should be very close if the ball is spherical.

4. The variable *Wgt* indicates the weight of the ball.

5. The variables *Dw* and *Dd* are measurements of the dimples. Dw indicates the dimple width and Dd indicates the depth.

6. The variable *Head* is the speed on the ball when it is hit.

7. The variable *Temp* and is the air temperature measured in degrees Fahrenheit.

8. The variable *Carry* indicates the distance from the point the ball was hit to the point where it hit the ground, measured in yards.

9. The variable *Tot Dist* indicates the total distance the ball has traveled. It is equal to the variable Carry plus the distance the ball rolled after hitting the ground.

10. The *Date* is 8/20 for all 72 observations.

11. The variable *Time* shows the time of day when the observation was recorded. There are 3 different time periods:

8:15 AM -- Ball#   1-24
8:45 AM -- Ball# 25-48
9:15 AM -- Ball# 49-72

> *To open a file choose Open Worksheet from the File Menu Remember to resave the file immediately with a slightly different name.*

Read in the datafile named CHAP5.MTW using the commands described in Chapter 3. Remember to create a working version of the datafile by resaving the file with a slightly different name.

Section 5.4 How to Create a Histogram in Minitab

> *Choose Graph>Histogram*

In this section you will learn how to create frequency histograms, relative frequency histograms and manage graph windows

Section 5.4.1 Creating a Frequency Histogram

This section focuses on the basics of creating a frequency histogram using Minitab. Subsequent sections cover some of the more advanced features available in Minitab which pertain to histograms.

Suppose you want to generate a histogram for the variable *Carry* for all 72 observations. From the menu bar click on **Graph**. The pull down menu shown in Figure 5.4 will appear.

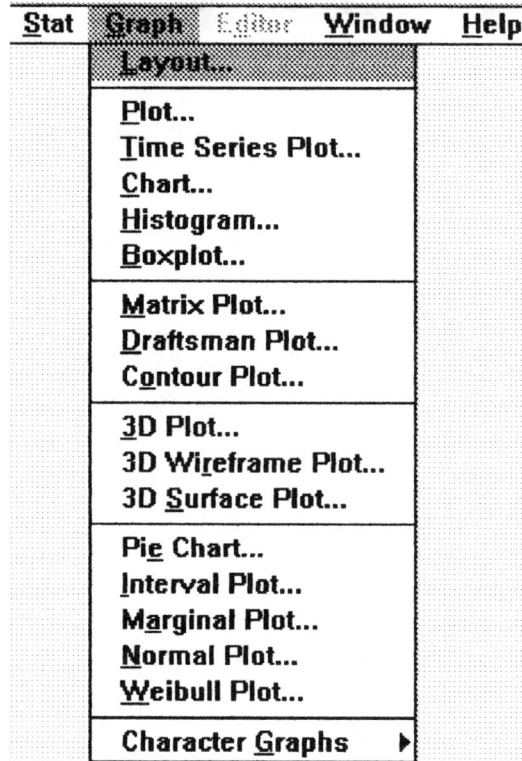

Stat	Graph	Editor	Window	Help

Layout...

Plot...
Time Series Plot...
Chart...
Histogram...
Boxplot...

Matrix Plot...
Draftsman Plot...
Contour Plot...

3D Plot...
3D Wireframe Plot...
3D Surface Plot...

Pie Chart...
Interval Plot...
Marginal Plot...
Normal Plot...
Weibull Plot...

Character Graphs   ▶

Figure 5.4 The Graph Menu

Choose **Histogram** from the Graph menu. You will see the Histogram Dialog box shown in Figure 5. 5 .

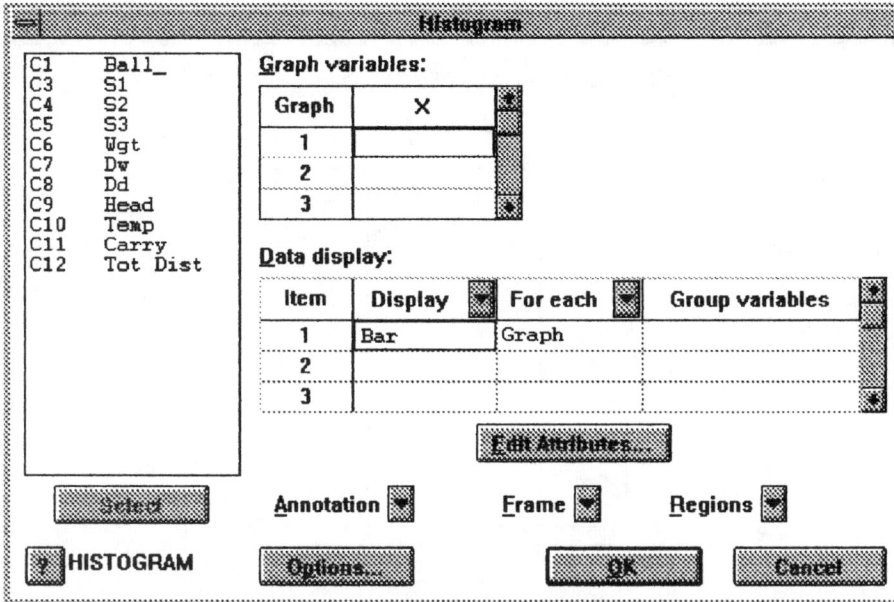

Figure 5.5   Histogram Dialog Box

Notice that Histogram dialog box has two sections that require input. The first is the **Graph** variables which is where you specify which variables you want to graph and the second is **Data display** which is where you specify what type of chart you want to use.

For now just specify the minimum information needed to create a histogram. The only thing you must do is tell Minitab what variable you wish to use in the histogram. In the dialog box you see a list of the possible variables from this worksheet. Select *Carry* by double clicking on the variable name *Carry*. You will see the variable name Carry appear in the **Graph variables** dialog box.

Double click on Carry *to select it for the* **X** *box*

The default type of chart is a bar chart so you don't need to do anything in the **Data display** dialog box. Now click **OK**. After a few seconds, the histogram appears in its own window. The histogram is shown in Figure 5.6

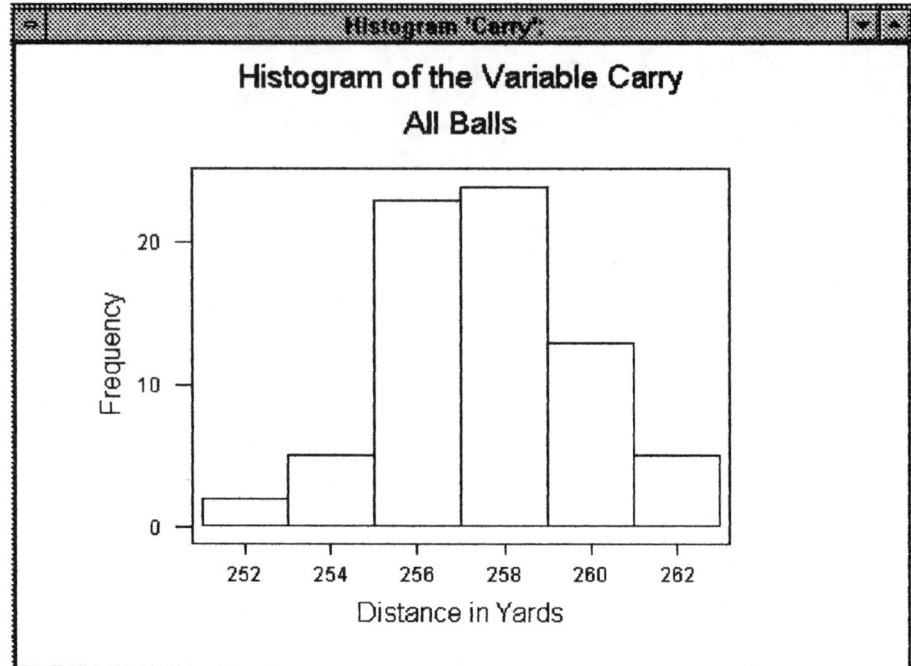

Figure 5.6 Frequency Histogram of *Carry* for all Observations

Remember to use the Annotate pop-up menu button to add the title to the graph. Use the Frame pop-up menu to label the axis.

You may switch back to the data window or any other window you wish by choosing Window and selecting the window you wish to make active.

Exercise 1. What does the histogram tell you about the distribution of how far the balls carry? How far do most of the balls carry?

Section 5.4.2 The Basic Relative Frequency Histogram

Notice that the vertical axis of the histogram shown in Figure 5.6 shows the frequency or raw count for each of the bars in the graph. Sometimes it is more helpful to have the relative frequency or percent displayed on the vertical axis. Minitab allows you to do this easily by clicking on the **Options** button in the Histogram Dialog box. When you do that you will see the **Histogram Options** dialog box shown in Figure 5.7

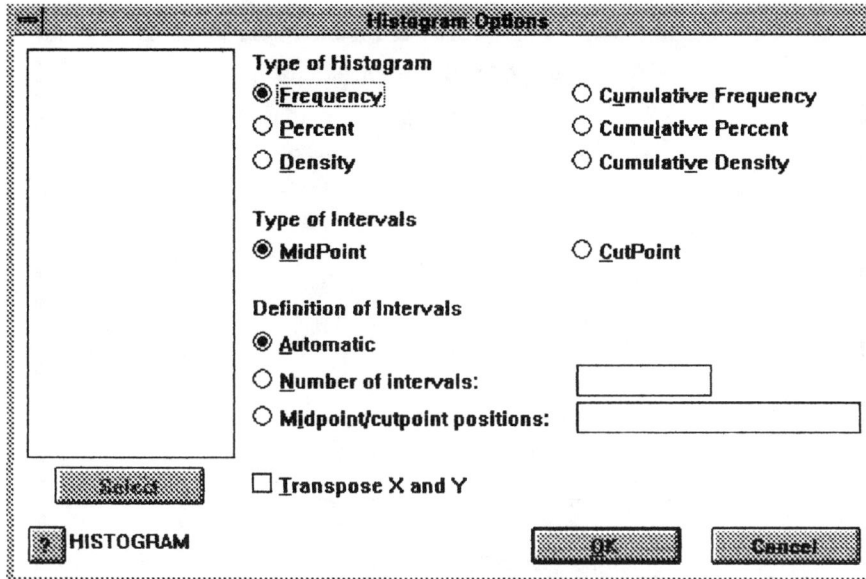

Figure 5.7 Histogram Options

The dialog box shown in Figure 5.7 shows the default
selections for the histogram options. In order to display percents
along the vertical axis simply choose ~~Frequency~~ Percent as the **Type of
Histogram** and click on **OK.** This will return you to the main
Histogram dialog box and then click on **OK** in this box and the
frequency histogram shown in Figure 5.8 will be displayed in a graph
window.

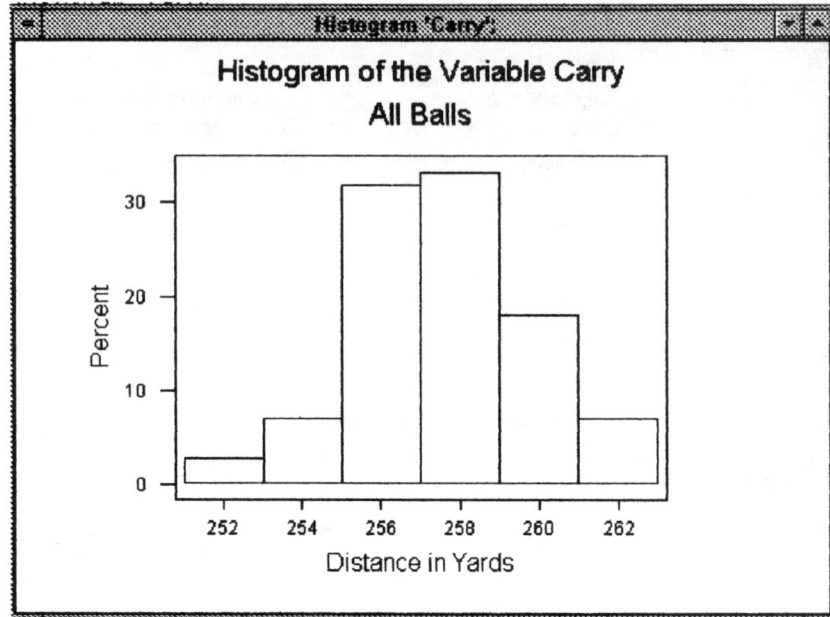

Figure 5.8 Relative Frequency Histogram of
***Carry* for all observations**

As you can see by looking at Figure 5.7 there are 6 different
types of histograms which you can create: frequency, percent, density,
cumulative frequency, cumulative percent and cumulative density. For
most data sets, constructing either the frequency or percent histograms
is the most useful.

Exercise 2. Create a frequency histogram and a relative frequency
histogram for the variable *TotDist* for all observations. How do these
compare to the corresponding histograms for the variable *Carry*?

Section 5.4.3 Using Graph Windows

Minitab can have up to 15 Graph Windows open.

You should have noticed that each histogram appears in what
is called a Graph window. A Graph window contains a single graph.
It is helpful to know a little bit about how to work with these graph
windows.

Closing a Graph Window

Minitab can have up to 15 Graph windows open at a time.
Generally you will not run into any difficulty with this limitation as you
will not often need to have that many graph windows open

simultaneously. You can close an individual Graph window by choosing **Close** from the control menu for the graph. You may close all Graph windows by choosing **Windows>Close All Graphs**. When you close a Graph window it is no longer available to you for viewing or printing unless you recreate the graph. Thus, if you are planning to use the graph at some later time during the session either leave it open or save it.

You may wish to print the contents of the Graph window before closing it. To do so make sure the Graph window is active then choose **File>Print Window** , then click **OK** . Finally, you can perform many Graph window functions with the **Window>Manage Graphs** dialog. The dialog box is shown in Figure 5.9

Printing a graph

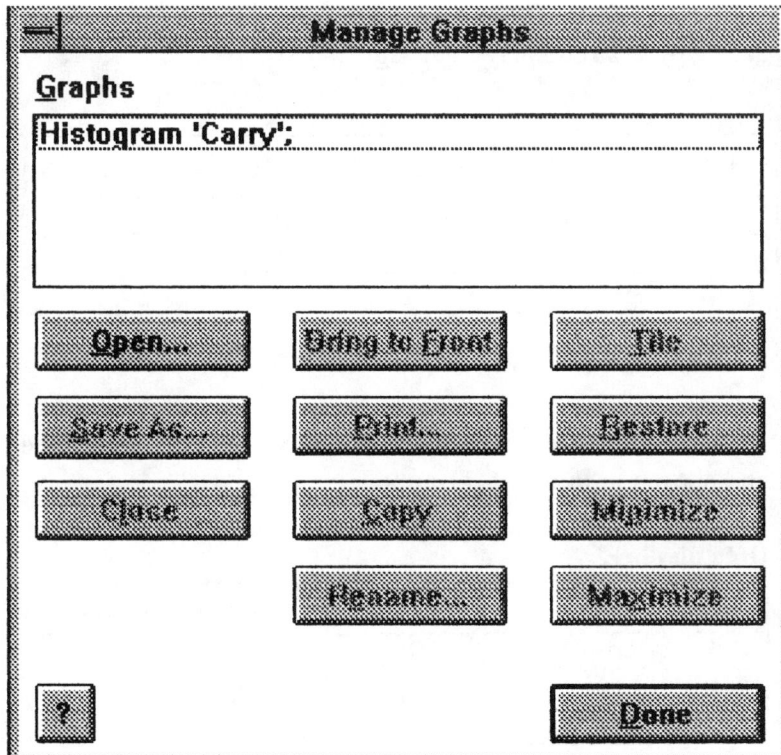

Figure 5.9 Manage Graphs Dialog Box

You may select a single graph window by clicking on the graph name. The Graph manager then allows you to perform a variety of functions such as: Close, Print, etc. Select the appropriate function by clicking on it.

Section 5.5 Creating Histograms for Subsets of a Variable

The major objective in the analysis of this data set is to determine which of the ball designs is "best". If we examine the histogram in Figure 5.6 or Figure 5.8 we can not compare the variable *Carry* for the two different ball designs because we have displayed all 72 values of the variable *Carry* in one histogram. In order to compare the *Carry* of the ball for the two different ball designs we must somehow distinguish the data values in the graph

Section 5.5.1 Using the Group Feature in Creating a Histogram

Minitab allows you to do this easily by specifying a **Group Variable** in the **Data display** dialog box of the Histogram dialog box. This dialog box is redisplayed in Figure 5.10.

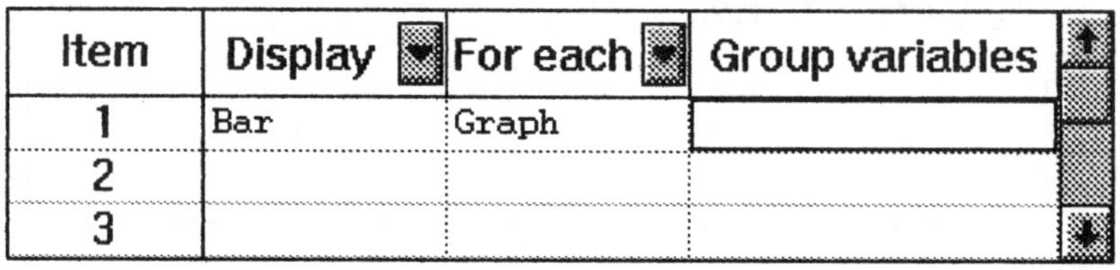

Item	Display	For each	Group variables	
1	Bar	Graph		
2				
3				

Figure 5.10 The Data display Dialog Box

If you click on the arrow button next to "**For each**", the pop-up menu shown in Figure 5.11 appears.

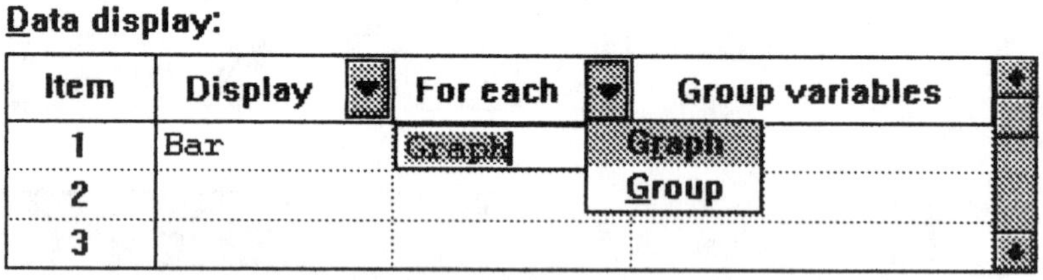

Data display:

Item	Display	For each	Group variables	
1	Bar	Graph	Graph	
2			Group	
3				

Figure 5.11 The "For each" Pop-up Menu

From this menu select **Group**. Now you must specify the variable which will identify the groups. In our case, we wish to group by the Model number of the ball. Move the cursor to the location under the words **Group variables** and double click on the variable named *Model_* shown in the list of variables. The completed **Data display** dialog box now looks like Figure 5.12.

Item	Display ▾	For each ▾	Group variables
1	Bar	Group	'Model _'
2			
3			

Figure 5.12 The Completed Data display dialog box

Then click **OK** and you will see the histogram shown in Figure 5.13

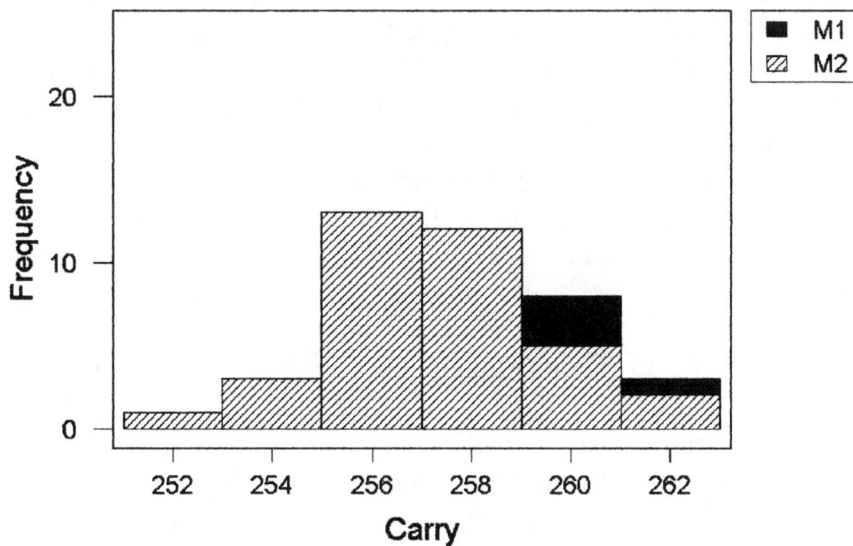

Figure 5.13 Histogram of Carry grouped by Model#

You can change the color and style of the bars by clicking on Edit Attributes in the main Histogram dialog box

Figure 5.12 seems to tell us that there are only a few observations of the variable *Carry* for the M1 ball design. These are the areas of the histogram displayed in solid black. However, we know from the description of the data that there were an equal number of balls tested for each of the two ball designs. The problem here is that the values of the variable *Carry* are very similar for both models.

Thus when Minitab creates this histogram it firsts displays the values for the variable *Carry* for the M1 ball design in solid black. Then it overlays the graph of the variable *Carry* for the M2 ball design using a cross hatched display. Thus, most of the values for the M1 ball design are covered over by those for the M2 ball design. The only values for the M1 ball design which remain visible (in solid black) are the ones which do not get covered over by the M2 ball design values.

Section 5.5.2 How to Convert Alpha Data to Numeric Data in Minitab

In the previous section we saw that the "group" feature in Minitab is only useful if the values in the 2 groups do not overlap. In order to see the behavior of the variable *Carry* for the two ball designs we will have to create two different histograms and compare them. However the values for the variable *Carry* for ball design M1 are not all next to each other in the column labeled *Carry*. In particular they are in rows 1-12, 25-36, and 49-60. What we need to do is copy the values for the variable *Carry* which correspond to ball design M1 into a blank column. There are several ways to do this in Minitab. One way to accomplish this is to first covert the column labeled *Model* from alpha data to numeric data. Since many of the Minitab options only work on numeric data, it is useful to learn how to do this at this point.

Creating a conversion table

The first step in converting a column of alpha data to numeric data is to set up a conversion table. The conversion table will tell Minitab what numeric value to assign to each of the possibilities in the column with alpha data. Be careful to type the entries in the conversion table exactly as they appear in the column you are trying to convert. You may type the conversion table in any blank space in the spreadsheet.

In our case we will convert M1 to 1 and M2 to 2. The conversion table which has been entered into the spreadsheet is shown in Figure 5.14.

Conversion Table

C25-A	C26
M1	1
M2	2

Figure 5.14 Conversion Table for Model

Now we are ready to convert the column C2-A. From the menu bar click on **Manip**. The pull-down menu shown in Figure 5.15 will appear.

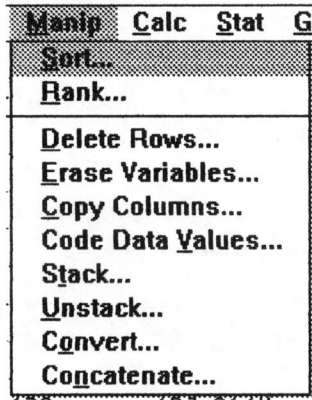

Figure 5.15 The Manip Menu

From this menu select **Convert** and you will see the dialog box shown in Figure 5.16

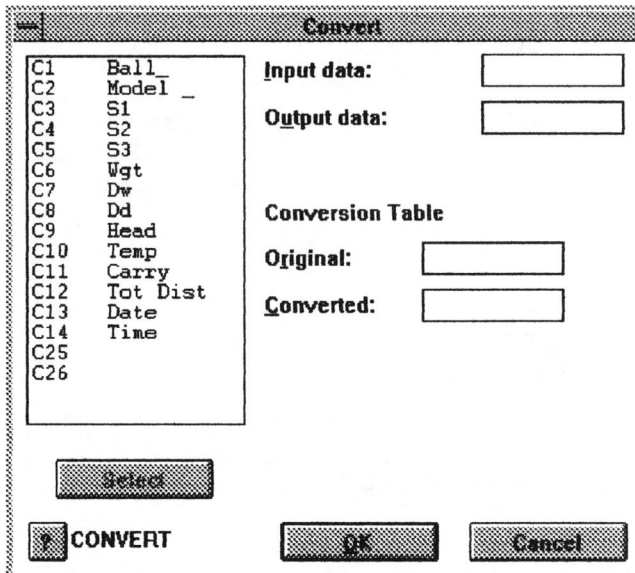

Figure 5.16 The Convert Dialog Box

To complete this dialog you must specify the **Input data**, the **Output data**, and the location of the **Conversion Table**. Move the cursor to the box labeled **Input data** and click on the box. The Input data is the column you wish to convert. The list of possibilities appears in the rectangle at the left of the dialog box. Double click on C2 *Model* to select it as the Input data. The Output data should be the column number of an empty column which is where Minitab will place the converted Model numbers. Finally, indicate the column locations of the conversion table. In box labeled **Original** in the **Conversion Table** section of the dialog box, enter the column number of the first part of the conversion table you created. This is the list of all the different possibilities of the variable you wish to convert. In the box labeled Converted enter the column number of the second part of the conversion table you created. This is the list of the numeric values to be assigned to each alpha value. The completed Convert Dialog Box for our example is shown in Figure 5.17

*Completed Convert
Dialog Box*

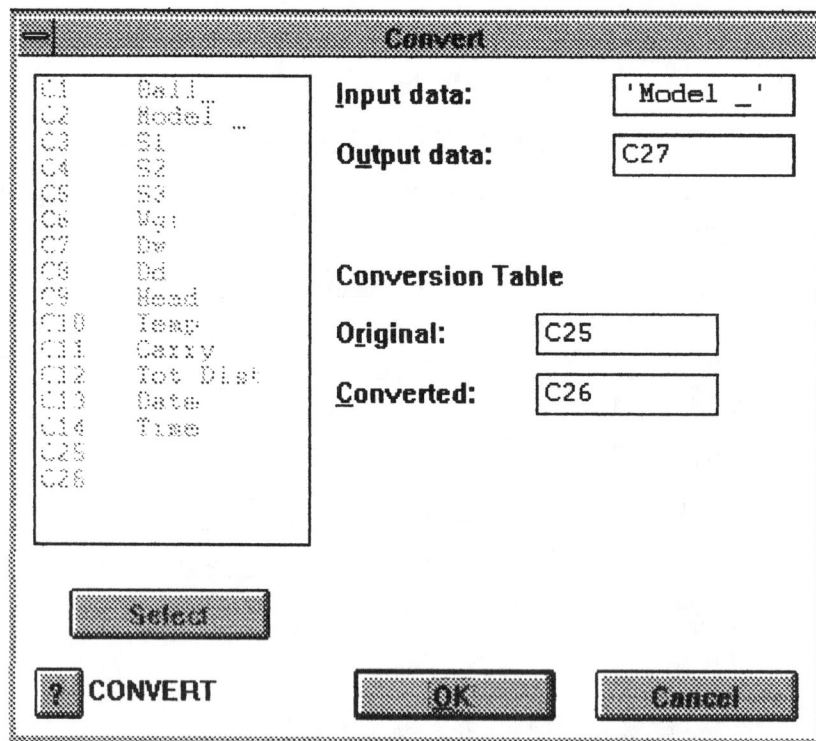

Figure 5.17 Complete Conversion Dialog Box

Select **O̲K** and you will see the converted alpha values in the spreadsheet in the column you specified for the Output data. In our case this is column C27 which we have now labeled *ModelNo*. The first few rows of this column are shown in Figure 5.18

```
         C27
       ModelNo
             1
             1
             1
```

**Figure 5.18 A few rows of the converted model
numbers**

We are now ready to copy the values of *Carry* for the ball with ModelNo 1 into an empty column of the spreadsheet. This will then be repeated for those values of *Carry* corresponding to the balls with ModelNo2 in order to create two separate histograms. This is explained in the next section.

Section 5.5.3 Copying Subsets of Data from a Column

Before doing the copying it would be best to label two columns in the spreadsheet. One will be labeled Carry - M1 and the other Carry - M2. These are empty columns at the moment but will be used to store the values of Carry from the copy procedure.

Manip>CopyColumns

From the menu bar choose **Manip**. You will see the pull-down menu shown in Figure 5.15. Choose **C̲opy Columns** and you will see the Copy dialog box shown in Figure 5.19

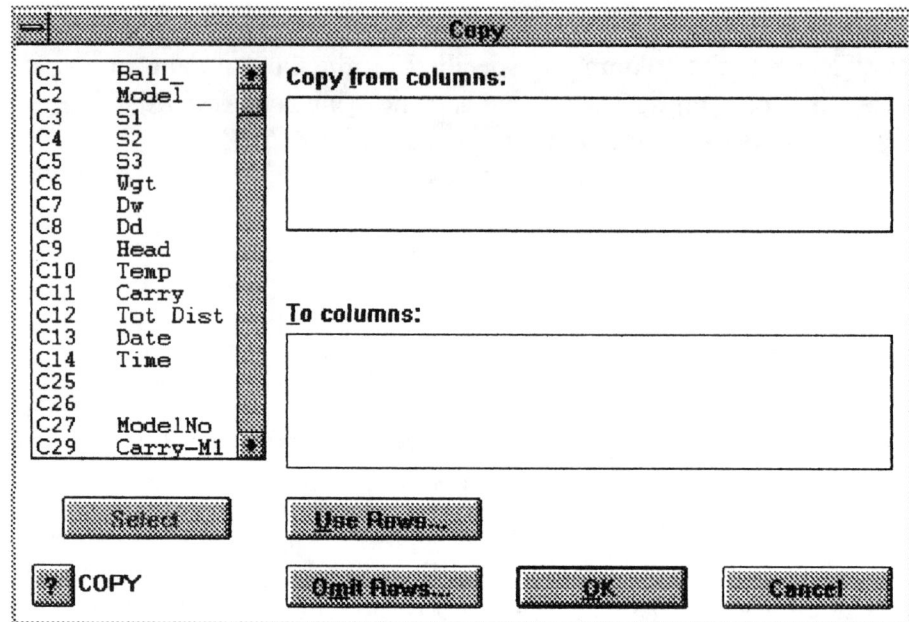

Figure 5.19 Copy Dialog Box

As we have seen before, you can specify the column you wish to copy by positioning the cursor in the **Copy from columns** dialog box and then double clicking on the variable name shown in the rectangular box. In this case we wish to **Copy from column** C11 which is the *Carry* column. Moving the cursor to the **To columns** dialog box and double clicking on column C29 which is labeled Carry-M1. Next click on the box which says Use Rows and you will see a dialog box similar to that shown in Figure 5.20.

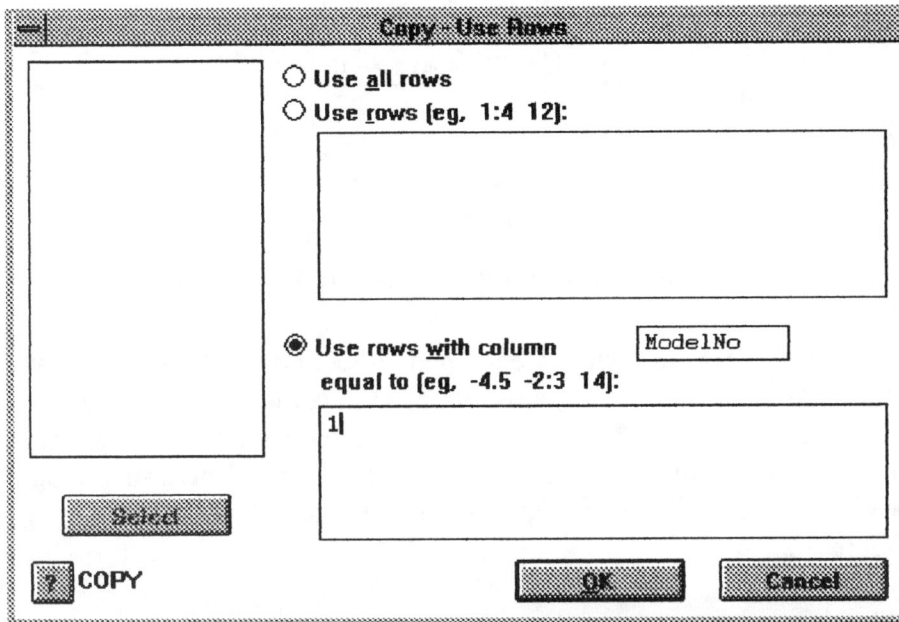

Figure 5.20 Copy Use Rows Dialog Box

The dialog box shown in Figure 5.20 has already been completed for our purposes. In order to get your dialog box to look like this first click in the circle next to the words **Use rows with column.** Then position the cursor in the blank box next to those rows and double click on the column which has the converted Model numbers; called ModelNo. Finally tell Minitab that you wish to copy the values in the Carry column to the column labeled Carry-M1 using those rows with ModelNo **equal to** 1 by typing a 1 in the box under the words **equal to**. Click **OK** and then click **OK** again in the Copy Dialog box and you will see the values of the variable carry for the M1 ball model copied into the column so labeled!

Exercise 3. Repeat this process to copy the values in the Carry column to the column labeled Carry-M2 using those rows with ModelNo equal to 2.

When you have completed Exercise 3 your spreadsheet should have two columns similar to the ones shown in Figure 5.21

C29	C30
Carry-M1	Carry-M2
257	256
255	255
256	258
255	257

Figure 5.21 Carry for each of the Ball Designs

Using the procedure we learned in Section 5.4.1, we can now create two frequency histograms; one for the column Carry-M1 and a second one for Carry- M2. Note that you can create two histograms at a time by specifying two **Graph** variables in the Histogram dialog box. The relevant portion of the Histogram dialog box is shown in Figure 5.22

Graph variables:

Graph	X	
1	'Carry-M1'	
2	'Carry-M2'	
3		

**Figure 5.22 Specify two Graph Variables to
generate two separate histograms**

Once you click on **OK** in the Histogram dialog box you will see the histogram for Carry-M1 fly by you on the screen and you will be left with the display of the histogram for Carry-M2 open in the Graph window. Most likely we would like to look at these two graphs at the same time. To do this choose **Window** from the menu bar and select **Tile** from this pull-down menu. This will cause all of your open windows to appear on the screen in a tile fashion. Close all the windows except for the two histogram you wish to view by clicking on the ▨ button in the upper right hand corner of each window. Now you can move the remaining two graphs anywhere you like by clicking on the title bar of the graph and dragging them to the position you like on the screen.

You should be able to get your screen to look like Figure 5.23

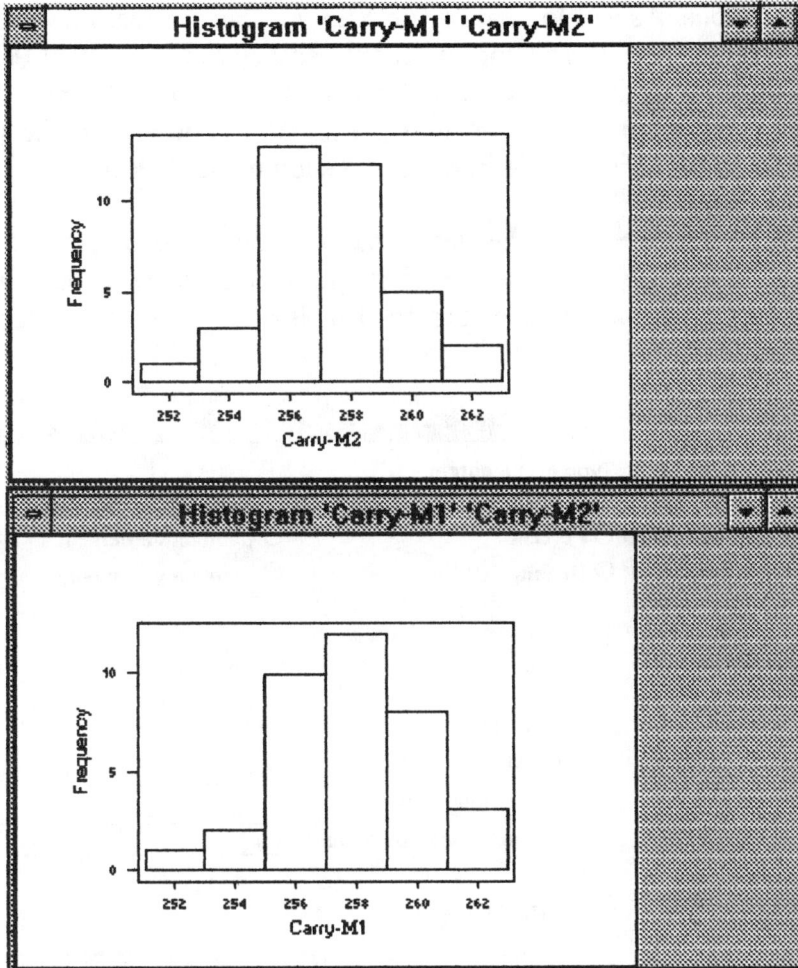

Figure 5.23 Histograms of Carry-M1 and Carry-M2

Exercise 4. Compare the histograms shown in Figure 5.23. What can you conclude about the behavior of the variable *Carry* for the two different ball designs?

Section 5.5.4 Histogram Options Useful for Comparing Histograms

When you are comparing histograms you need to be sure that you are making a fair comparison. That is you would really like to guarantee that the class intervals shown along the X-axis are the same in both graphs. As you look at Figure 5.23 you can see that they are in fact the same. This was a happy accident because of the fact that the data values for both ball designs are similar. This will not always happen without some help from us and Minitab. In order to make sure that your histograms are comparable you should use click on

the [Options...] from the Histogram dialog box (see Figure 5.5). When you do this you will see the Histogram Options dialog box which we have looked at once before. This is redisplayed in Figure 5.24

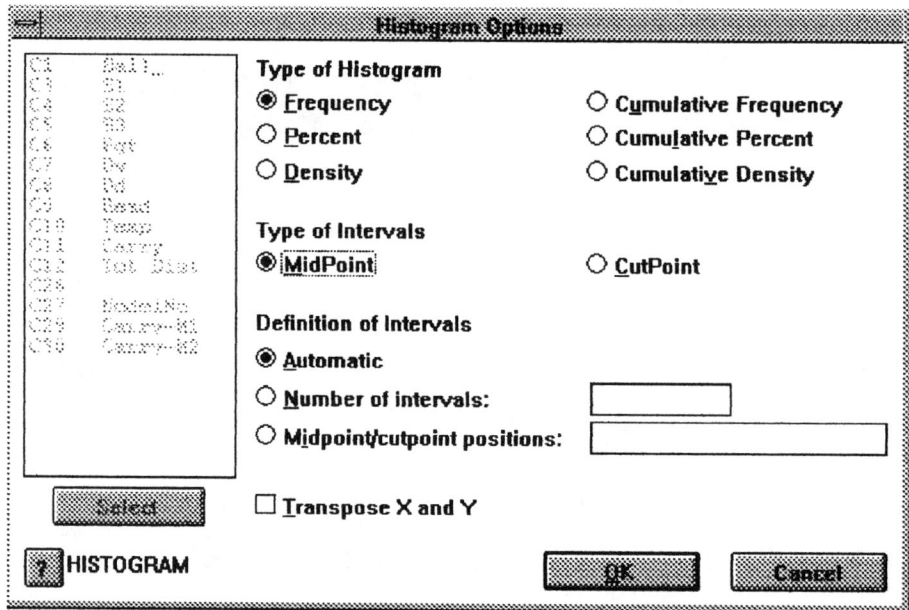

Using Cutpoints instead of Midpoints

Figure 5.24 Histogram Options dialog box

This time we are going to look at the **Type of Intervals** and the **Definition of Intervals**. All of the histograms we have looked at have had the tick marks appearing at the midpoint of the intervals. This is the default used by Minitab. It is often easier to read the graph with the tick marks appearing at the end of the intervals. Minitab will do this for you when you click on the circle next to the word **CutPoint** under **Type of Intervals.**

It is also helpful to be able to specify how many intervals you wish to have in the histogram and/or the values for the cutpoints. For example we may wish to specify that the first cutpoint be 250 and the last cutpoint be 270 and that the cutpoints increment by 4. To accomplish this we would click on the circle next to the words **Midpoint/cutpoint positions** and in the corresponding box we would type 250:270/4. Thus the histogram options would look like Figure 5.25

```
Type of Intervals
 ○ MidPoint                    ● CutPoint

Definition of Intervals
 ○ Automatic
 ○ Number of intervals:        [            ]
 ● Midpoint/cutpoint positions: [ 250:270/4        ]
```

**Figure 5.25 Histogram Options to Control Value
of Cutpoints**

The resulting histograms are shown in Figure 5.26

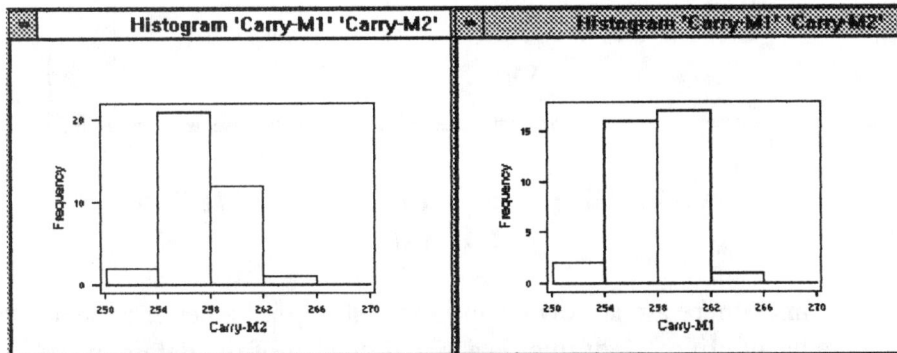

**Figure 5.26 Histogram of Carry-M1 and Carry-
M2 with Cutpoints Defined**

Section 5.6 Graphical Displays for Small Data Sets

Sometimes you need to create a graphical display of only a small number of points. For example suppose we wanted to display the values for the variable *Carry* for the M1 ball design taken at 8:15AM. There are 12 observations. In order to create a histogram of this subset of data we would need to copy those 12 values to a new column and create a histogram of that new column. Let's see how that would work. In this case we do not need to convert any data values since the 12 values are next to each other in the original *Carry* column. From the menu bar Choose **Edit>Copy Cells** to copy the values in rows 1-12 for the variable *Carry* to a new column and create the histogram from that column. The histogram is shown in Figure 5.27

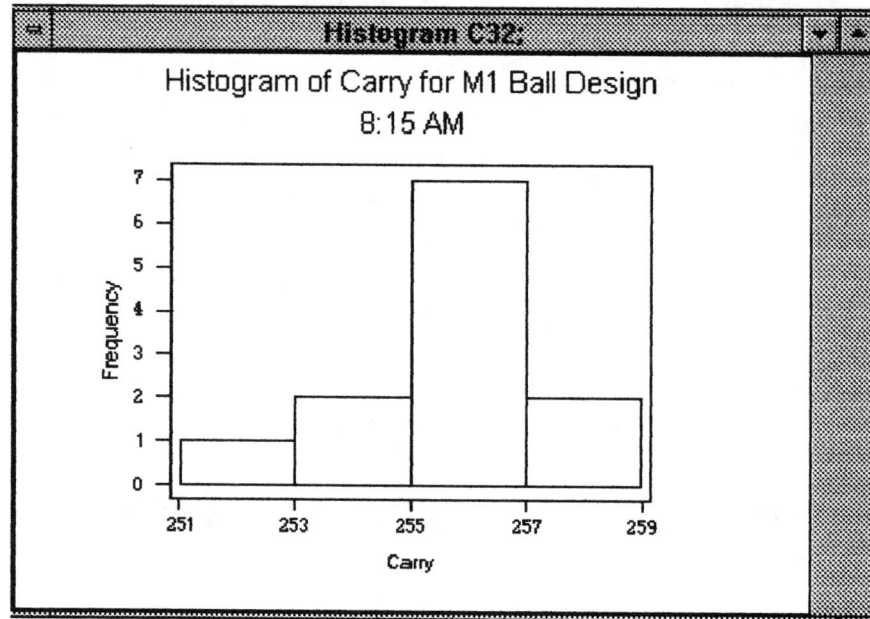

**Figure 5.27 Histogram of *Carry* for the M1 Ball
at 8:15AM**

Since there are so few observations it might be better to view the data as a dotplot. A dotplot is a visual display of the data that shows a dot for each of the data values.

To create a dotplot of the 12 *Carry* values currently in Column C32 choose **Graph>Character Plots** from the menu bar. When you do that the pop-up menu shown in Figure 5.28 will appear.

Set Options...

Histogram...
Boxplot...
Dotplot...
Stem-and-Leaf...

Scatter Plot...
Multiple Scatter Plot...

Time Series Plot...

Grid...
Contour...
Pseudo 3-D Plot...

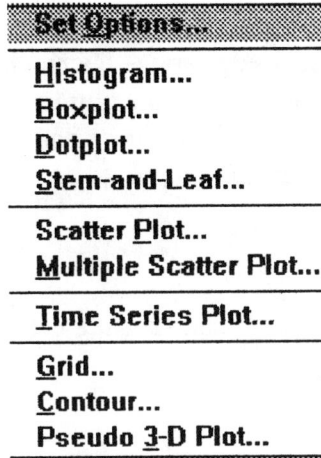

Figure 5.28 Character Plot Pop-up Menu

From this menu pick **Dotplot.** The dotplot will be displayed in the Session Window and looks like the one shown in Figure 5.29

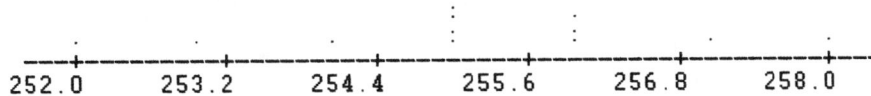

```
                         .               .
         .          .    .    .        .          .          .
    --+---------+---------+---------+---------+---------+---------+---
    252.0      253.2      254.4      255.6      256.8      258.0
```

**Figure 5.29 Dotplot of *Carry* for M1 Ball Design
at 8:15 AM**

You can see that each and every data value is visible in the dotplot. Thus it is only useful if there is a small number of data points to be displayed.

You may have noticed an option for Histogram in the Pop-up menu shown in Figure 5.28. Do **not** use this option as it will create a character histogram for you which uses character symbols to display the height of the bars in the histogram. It is a carry over from "old technology".

Section 5.7 Investigative Exercises

In the following exercises you are asked to create a variety of different graphs which will display various aspects of the golf ball datafile. The graphs are then to be used to draw conclusions about the datafile and make recommendations to management.

DO NOT treat the individual exercises or even the individual components of an exercises as isolated graphs. Remember the reason you are visually displaying the data is to enable you to "see" the information contained within the data. Look for trends, patterns, similarities, and differences between the graphs.

1. Construct a histogram for:

Note: you do not have to save the graphs if you print them as you create them.

a) The variable *Carry* for the M1 ball design for the 8:15 time period. Save the graph as G51a.

b) The variable *Carry* for the M1 ball design for the 8:45 time period. Save the graph as G51b.

c) The variable *Carry* for the M1 ball design for the 9:15 time period. Save the graph as G51c.

d) The variable *Carry* for the M1 ball design for all three time periods combined. Save the graph as G51d.

e) Does the variable *Carry* exhibit generally the same behavior in each time period?

f) What, if any, differences did you see?

2. Generate four histograms similar to those constructed in exercise 1 using the M2 ball design. Save the graphs as G52a, G52b, G52c, and G 52d.

a) Graph G52a:

b) Graph G52b:

c) Graph G52c:

d) Graph G52d:

e) Does the variable *Carry* exhibit generally the same behavior in each time period?

f) What, if any, differences did you see?

3. Compare the graphs from exercise 1 with those generated in exercise 2. What can you conclude about the difference, if any, between the two ball designs with regard to how far the ball carries?

4. Generate histograms to display the total distance the M1 ball traveled during each of the three time periods individually and combined. Save the graphs as G54a, G54b, G54c and G54d.

a) Graph G54a:

b) Graph G54b:

c) Graph G54c:

d) Graph G54d:

e) Does the variable *Tot Dist* exhibit generally the same behavior in each time period?

f) What, if any, differences did you see?

5. Generate four histograms similar to those constructed in exercise 4. Use the M2 ball design. Save the graphs as G55a, G55b, G55c, and G55d.

a) Graph G55a:

b) Graph G55b:

c) Graph G55c:

d) Graph G55d:

e) Does the variable *Tot Dist* exhibit generally the same behavior in each time period?

f) What, if any, differences did you see?

6. Compare the graphs generated in exercise 4 with those generated in exercise 5. What, if any, differences are there in the total distance that the ball flies for the two ball designs?

7. Experiment with a different cutpoint positions. For instance, try creating a graph of the variable *Tot dist* for the M1 golf ball with the cutpoints 1 yard apart, then using 5 apart, and then 10 apart. Compare these 3 graphs.

a) Which graph gives management the most informative view of the data? Why did you select this one?

b) What happens to the graph if your cutpoints are too close to each other (resulting in too many class intervals)?

c) What happens to the graph if your cutpoint positions are too far apart (resulting in too few class intervals)?

8. Construct a histogram of the variable *Ball #*.

a) What does it look like?

b) Why should you have expected it to look like this?

c) Why is it a useless graph?

9. Change the values of *Carry* and *Tot Dist* for Ball# 1 to 357 and 370 respectively.

a) Create a histogram of the variable *Carry* for the M1 ball design for the time period 8:15.

b) Compare this to the graph saved as G51a. What has happened to the histogram as a result of this unusual data point?

10. Visually examine the variable *Temp* in the spreadsheet.

a) What do you notice?

b) Construct a histogram of all 72 temperature values.

c) Does the histogram show what you noticed about the temperatures?

d) Create a different way of displaying the temperatures which will in fact display what you noticed when you visually examined the temperature values.

11. Of what use is it to have the variable *Date* when they are all the same?

12. Look at the variables *Carry* and *Tot Dist* for ball numbers 64 and 66.

a) What do you notice?

b) Assuming you have correctly controlled for all the factors, what is causing the variation in the total distance even though the ball carried the same yardage in both cases?

13. Construct whatever histograms you need to examine whether any of the controlled factors, internal of external, were really not held constant during the test.

14. On the basis of your analysis of the histograms you created, which of these two ball designs would you recommend to management. Support your answer with the appropriate graphs that illustrate your points.

Chapter 6   "Golf Ball Design"
Numerical Descriptors

Section 6.1 Overview

In the last two chapters you have studied graphical methods for displaying data. Although these methods provide a visual picture of the data, they do not provide any numerical, summary information about the data. This chapter will focus on the most commonly used summary statistics. These are also called numerical descriptors or descriptive statistics. You will see how to use **MEASURES OF THE MIDDLE** and **MEASURES OF DISPERSION** to describe a large dataset.

Statistical Objectives: After reading this chapter and doing the exercises a student will:
- Know how to use the mean as a measure of the middle and understand its limitations.
- Know how the median differs from the mean and in what situations it is a better measure of the middle.
- Know that by comparing the mean and the median you can tell the general shape of the distribution.
- Know the meaning of the variance and the standard deviation.
- Know that the z-score tells you how many standard deviations the data point is from the mean.
- Know how to use the empirical rule.

Section 6.2 Problem statement

We return to the dataset from Chapter 5 concerning two different golf ball designs. Recollect we introduced a manufacturer interested in product design and quality improvement who was collecting a lot of data on the performance of its golf balls. The actual variables of interest to which the firm was directing its research were flight and total distance the ball carries. W. Edwards Deming, a founding statistician of the modern quality assurance movement, has commented that raw data itself is meaningless without the analyst having strong skills in understanding its most difficult feature, random variability. Developing the skills of interpretation and prediction with sample data, Deming would say, are the keys to understanding and controlling product variability and to managing consistent product improvement. As you develop numerical descriptors to summarize the golfball datafile, keep Deming's insights in mind.

Section 6.3 Calculating numerical descriptors using Minitab

Use
File Open Worksheet
to read the file.

Section 6.3.1 Descriptive Statistics for the entire column of data

Read in the golfball datafile named CHAP5.MTW from your data disk. Remember to create a working version of the datafile by resaving the file with a slightly different name.

From the menu bar click on **Stat**. The pull down menu is shown in Figure 6.1

*Choose **Stat>Basic Statistics***

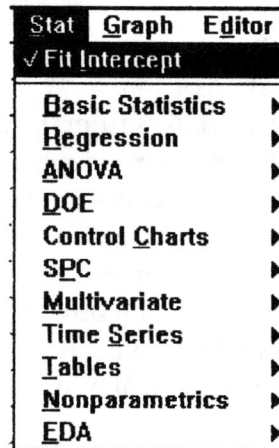

Stat	Graph	Editor
√ Fit Intercept		

Basic Statistics ▶
Regression ▶
ANOVA ▶
DOE ▶
Control Charts ▶
SPC ▶
Multivariate ▶
Time Series ▶
Tables ▶
Nonparametrics ▶
EDA ▶

Figure 6.1 Stat Pull down Menu

From this menu select **Basic Statistics** and the pop up menu shown in Figure 6.2 will appear.

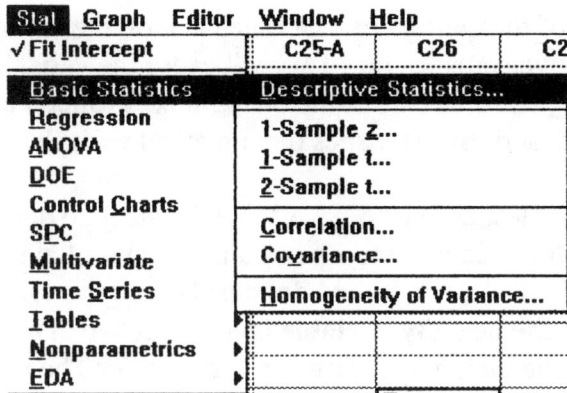

Figure 6.2 Basic Statistics Pop Up Menu

Select the option **Descriptive Statistics**. You will see the dialog box shown in Figure 6.3

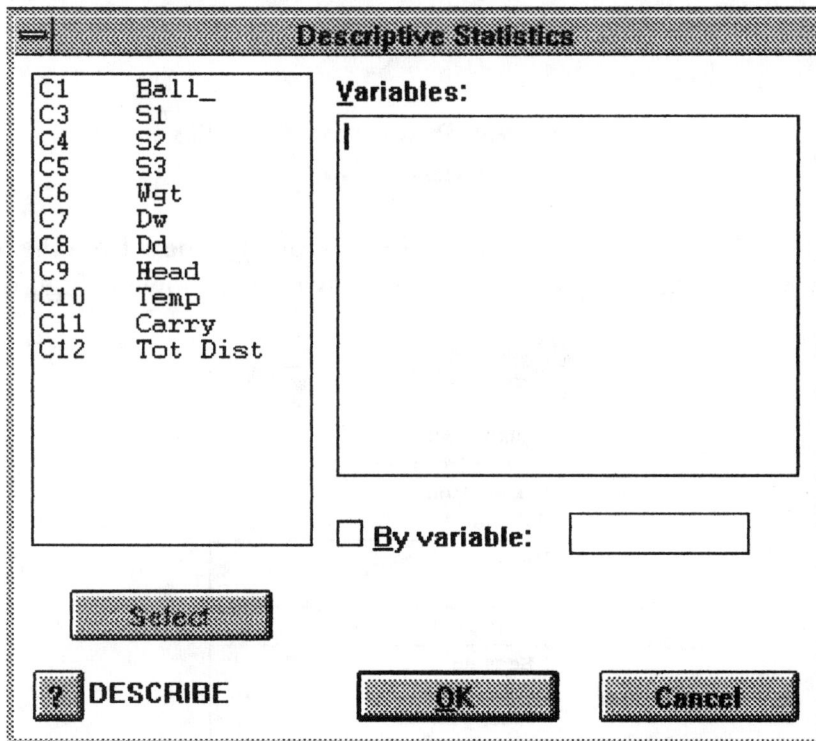

Figure 6.3 Descriptive Statistics Dialog Box

When you position the cursor in the **Variables** part of the dialog box, the list of variables that you can "describe" appears in the left hand side of the dialog box. Double click on the variable you wish to run descriptive statistics on. This will cause the variable name to appear in the **Variables** part of the dialog box. Notice that you can only perform descriptive statistics on numerical variables.

Descriptive Statistics for an entire column of data

The simplest case is to have Minitab calculate descriptive statistics for **all** the data in the a particular column. Let's use the variable *Carry*. Thus, you would double click on the variable *Carry* and make sure the box **By variable** is **not checked**. Then click the **QK** button in the dialog box shown in Figure 6.3. The results will appear in the Session window and your screen will look like Figure 6.4

*If you forget to uncheck the **By variable** box you will get this error message.*

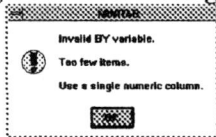

Descriptive Statistics

Variable	N	Mean	Median	TrMean	StDev	SEMean
Carry	72	257.10	257.00	257.12	2.33	0.27

Variable	Min	Max	Q1	Q3
Carry	251.00	262.00	255.00	258.75

Figure 6.4 Descriptive Statistics for the variable *Carry*

To return to the data window choose **Window** from the main menu bar. This will give you the pull down menu shown in Figure 6.5.

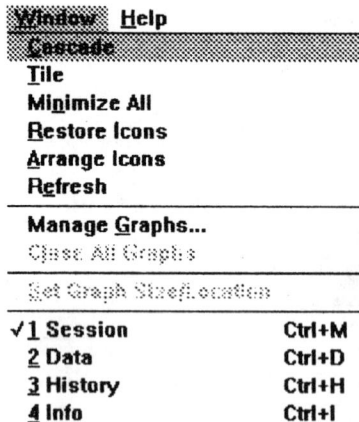

*Use **Ctrl+D** as a shortcut way to return to the Data Window.*

Figure 6.5   Window pull down menu

The checkmark next to the word Session indicates that you are in the Session window. Choose Data if you want to return to the Data window where we have been working.

Let's look at what Minitab has calculated for us.

Section 6.3.2 "Measures of the Middle" calculated by Minitab

Minitab generates two of the three commonly used measures of the middle of a set of data. These are the mean and the median. For the example data these values can be extracted from the screen shown in Figure 6.4. The mean of the variable *Carry* is equal to 257.1 and the median is equal to 257.0.. A third measure of the middle which is often quoted is the mode. Despite the fact that it is frequently quoted, the mode is not a very good measure of the middle, and is thus not calculated by Minitab.

Many times when data are analyzed the mean and the median are calculated and quoted but never really examined. Take a few minutes to *see* what information is actually contained in these two numerical descriptors.

Recall that the mean is found by adding up all the data values and dividing by the number of observations. Also remember that the median is found by sorting the data from smallest to largest and selecting the observations which falls right in the middle of the ranked data. Think about what will happen to the mean and the median if there are one or two really large numbers in the data set.

In order to see the effect of an outlier you need to look at what portion of the formulas will be impacted. In the case of the median, a large value does not have any impact because the median is simply the middle score. It does not matter how big or how small the extreme values are, the median will still be the same.

In the case of the mean, clearly a large observation will cause the mean to be higher because the sum of the data values will be higher.   Minitab will calculate a variety of other statistics, including the sum of the column. Choose **Calc** from the main menu bar. When you do this you will see the pull down menu shown in Figure 6.6

*Choose **Calc>Column** Statistics to **add up a column.***

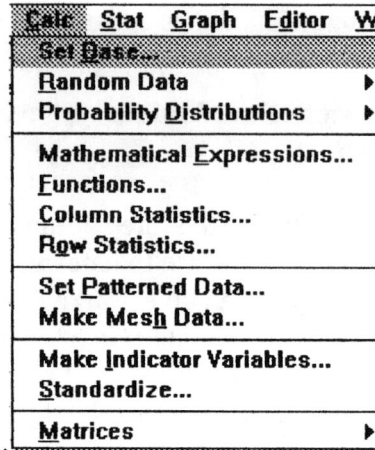

Figure 6.6 The Calc Menu

Choose **Column Statistics** from this menu and you will see the
Column Statistics dialog box shown in Figure 6.7

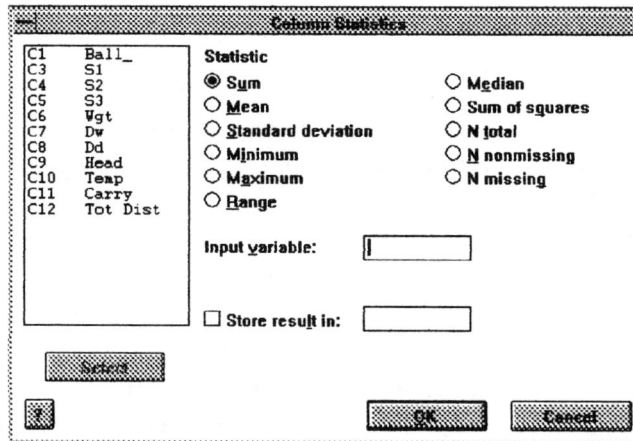

Figure 6.7 Column Statistics Dialog Box

Notice that this dialog box gives you the option to calculate a
variety of statistics on a column of data, including all the ones
generated by the **Stat>Basic Statistics>Descriptive Statistics**
command sequence. However you can only ask it to calculate one of
these statistics at a time. The advantage of using the **Stat>Basic
Statistics>Descriptive Statistics** command sequence is that you can
calculate all the descriptive statistics at once.

Returning to our problem, check the **Sum** box (it is by default checked) and double click on the variable *Carry* once your cursor is located in the **Input variable** section of the dialog box shown in Figure 6.7. Now click on the **OK** button. You will see the sum of the variable *Carry* shown in the Session window and it will look like Figure 6.8.

Column Sum

```
    Sum of Carry    =           18511
MTB > Sum 'Carry'.
```

Figure 6.8 Sum of the variable *Carry*

*Use **Ctrl D** to return to the Data Window*

You should see the sum 18511 displayed. When this is divided by the number of observations which is shown as N in Figure 6.4 , you get the mean value of 257.1.

Exercise 1: Change the value in the first row of the variable *Carry* to 387. Recalculate the column sum. What happens to the sum? Re-issue the command sequence **Stat>Basic Statistics>Descriptive Statistics**. What happens to the mean?

Exercise 2: What happened to the median when you changed the value in row 1 of the variable *Carry* ? Recall that it was 257 prior to the change. What is it now?

Although the mean changes quite a bit, the middle of the data has not changed at all. Thus, the mean can be misleading if what you are trying to communicate is the middle of the data values. Unusually large or small observations are called *outliers*. They are values which lie very far from the middle of the distribution. An outlier may result from transposing digits when recording or observing an observation. Even when there are no recording or observational errors, a data set may contain one or more valid measurements which, for one reason or another, differ markedly from the others in the set. As you have seen, these outliers can cause a distortion in the sample mean, $\bar{x}$. Therefore, isolating outliers should be one of the first steps in any data analysis. When there are outliers in the dataset, the median gives a more accurate description of the "middle" of the data.

*Remember the **mean** is sensitive to outliers while the **median is not.** Use the trimmed mean to eliminate the effect of outliers.*

Another way to avoid having the mean give a misleading picture is to use what is known as the trimmed mean. Typically a 5% trimmed mean is calculated. This involves eliminating the lowest 5% of the values and the highest 5% of the values and then calculating the mean of the remaining data. Minitab does this for you and it is shown as **Tr Mean** in Figure 6.4.

Tr Mean gives you the value of a 5% trimmed mean.

Exercise 3:. What is the trimmed mean of the variable *Carry*? What is the median? You should be able to see that by dropping the extreme values the mean is now a more accurate reflection of the true middle.

The next question you should ask is "How can I tell if an observation is an outlier?" The measures of dispersion and boxplots discussed in the next two sections will help you answer this question.

However, before leaving the measures of the middle you should notice that before you changed the value in row 1 of the variable *Carry*, the mean and the median were almost equal in value.

Exercise 4: Examine the descriptive statistics of Carry shown in Figure 6.4. Compare the mean and median values.

In general, they will not be equal but you should *always* compare their values. By comparing them you can get an initial clue as to whether or not there are any outliers. Secondly, by comparing the mean and the median you can get a sense of the shape of the data. Often times the mean and the median are quoted but you do not have access to the data directly and a graph may not be provided. In this case, a simple comparison of these two values will give you a general idea about the graph.

When the mean and the median are close in value then the underlying distribution is symmetric and often has a bell shape. If the mean is much smaller than the mean then the shape of the graph is skewed to the left. This means that there are a few extremely low values which are dragging the mean down while the median is unaffected by these low values. If the mean is much larger than the median then the graph is skewed to the right. This means there are a few large values which are artificially pulling the mean up while the median is a much smaller value. When you look at the numerical descriptors together with the histogram of the data you may also use the values of skewness and kurtosis to describe the behavior of the graph.

Exercise 5: Using the tools you learned in Chapter 5, construct a histogram of the variable *Carry*. Does the graph confirm what you expected?

In order to decide whether or not a particular statistical technique can be used, it is often important to know whether the distribution has a symmetric shape. This matter will be further discussed in Chapters 10-14.

Section 6.3.3 "Measures of Variability" calculated by Minitab

In the screen shown in Figure 6.4 you can find one measure of variability. It is known as the standard deviation and is displayed as **StDev**. For the example, the standard deviation of the variable *Carry* is shown as 2.33. Information is also provided on the minimum and maximum values of the variable. A quick subtraction of the minimum value (shown as **Min**) from the maximum value (shown as **Max**) gives you the Range of the data which is another frequently quoted measure of variablilty. In this case, the range of the data is 262-251=11.

Exercise 6: Be sure you can find these numbers in the output.

As was the case with the measures of the "middle", the measures of dispersion are often simply calculated and then not used because people do not really understand what information they contain. Take a closer look at each of these measures to get a better idea of what they really tell you about the data. Remember your job is to extract as much information from the data in order to be able to make informed decisions.

The range is the easiest measure of variability to calculate and is therefore often quoted. Extreme care should be exercised in drawing any conclusions about the data on the basis of the range. By its very definition, Range = maximum score - minimum score, it is highly sensitive to the extreme values in the dataset. By changing either the largest or smallest value or both, the range is quickly and often radically changed even though the amount of true variation in the data has not changed that much.

Exercise 7: Look at the output generated in Exercise 1, when you change one value from 257 to 387. Using the Max and Min values, calculate the new value for the Range.

You should notice that the range has changed rather dramatically from its original value of 11 although the only thing that has really changed is one data observation!   This characteristic makes the range a poor measure of dispersion.

The sample variance is often quoted but is useful primarily as a stepping stone to the standard deviation. By looking at the formula for the sample variance:

$$s^2 = \sum_{i=1}^{n} (x_i - \bar{x})^2 \ / \ (n-1)$$

you can easily see that the unit of measure on the variance is whatever unit of measure you were using squared. It is not particularly intuitive or helpful to think about dispersion in terms of units such as feet squared. Thus, the sample variance is most helpful simply as a step in getting the standard deviation. Thus Mintab does not print the sample variance but it could easily be calculated by squaring the standard deviation. The standard deviation, s, is simply the square root of the variance: $s = \sqrt{s^2}$.

*The **standard deviation** is the most useful measure of dispersion.*

The standard deviation is the most useful of all the measures of dispersion. Think about the standard deviation as a yardstick to be used to measure any and all differences in the dataset. For example, suppose the distance between a data value and the sample mean is 100 feet. You can not tell by simply looking at the value of 100 feet whether the data value is close to the mean or far away from the mean. This is because all differences are only meaningful when compared to the standard deviation. What you want to know is how to many standard deviations does 100 feet correspond. If the standard deviation was 50 then the difference of 100 translates to 100/50 or 2 standard deviations. However, if the standard deviation was 20 then the difference of 100 translates to 100/20 or 5 standard deviations.

What are these numbers 2 and 5 called? You know them as the z-score for the observation. Remember that the z-score is simply the difference between the observation and the sample mean divided by the standard deviation:

$$\text{z-score} = \frac{X - \bar{X}}{s}$$

The **empirical rule** tells you that 68% of all observations should fall within 1 standard deviation of the mean. This means that 68% of all observations should have a z-score between -1 and 1. It also tells you that about 95% of the observations should have a z-score between -2 and 2 . Finally it tells you that virtually all of the observations (99%) should fall within 3 standard deviations of the mean and thus have a z-score between -3 and 3. This implies that a z-score of 2 is reasonable but a z-score of 5 is clearly an outlier.

Exercise 8: For the variable *Carry* which you have been looking at, how many standard deviations away from the mean is the observation 255. Based on this calculation determine if it is an outlier?

Exercise 9: Calculate the z-score for Ball #15. Verify that it falls within one standard deviation of the mean by noticing that ball #15 carried a distance of 258 which is between the limits:

Mean + 1 standard deviation   and   Mean - 1 standard deviation.

Thus a z-score between -1 and 1 tells you that the observation falls within one standard deviation of the mean.

The final measure of dispersion which Minitab helps you calculate is the semi-interquartile range. This is found by taking the difference between the third quartile (shown as Q3) and the first quartile (shown as Q1) and dividing by 2. It is often called the Interquartile range: IQR = (Q3-Q1)/2. It is easiest to think about the IQR in conjunction with box plots which are discussed in Section 6.4.

Section 6.3.4   Using the By Variable feature of Descriptive Statistics

Using a grouping variable to create descriptive statistics

In Section 6.3.1 you learned how to use Minitab to find descriptive statistics for a complete column of data. It is often useful to be able to calculate descriptive statistics for subsets of a column of data. For example, with the variable *Carry* we would definitely want to look at the mean *Carry* for the different model types. Thus we want a mean of the variable *Carry* for model M1 and model M2. Minitab provides an easy way to do this as long as the variable that we are using to determine the subsets is a numeric variable. We ran into this issue in Chapter 5 when we wanted to create a histogram for *Carry* for each model. Recall that in that case we converted the alpha data stored in the column *Model_* to numeric data. (see section 5.5.2 for more details).

If you have completed Chapter 5 then you should already have a column in your worksheet with the converted model numbers. We called it *ModelNo*.   If not, create such a column using the **Manip>Convert** sequence of commands explained in Section 5.5.2. The newly created column will have a 1 for *Model _* M1 and a 2 for *Model_* M2. Type in the words *ModelNo* as a column label at the top of the column, if necessary.

Once this column has been created you are ready to tell Minitab to calculate descriptive statistics for the variable *Carry* by *ModelNo*. Choose **Stat>Basic Statistics>Descriptive Statistics** and select the variable *Carry* from the Variable List box as before. Now click on the **By variable** box and double click on *ModelNo* in the Variable List box to tell Minitab to create descriptive statistics by Model number. The completed dialog box is shown in Figure 6.9

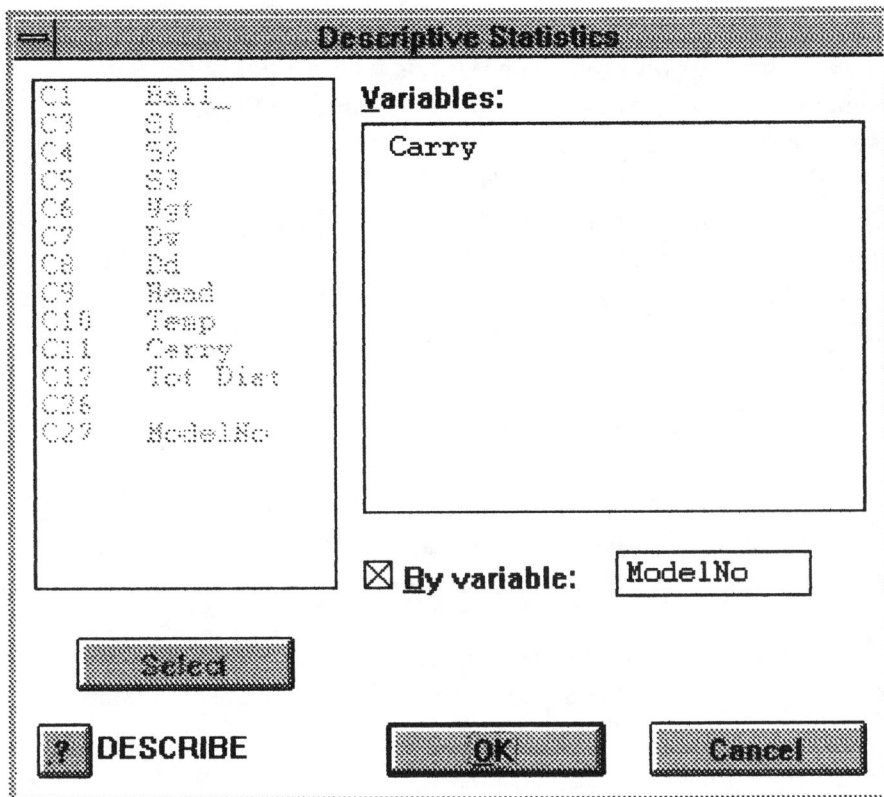

Figure 6.9 Completed Dialog Box for Descriptive Statistics of *Carry* by *Model No*

Click on the **OK** button and you will see the descriptive statistics for the variable *Carry* for each of the Model numbers. The output is displayed in the Session window and is shown in Figure 6.10.

Descriptive Statistics

Variable	ModelNo	N	Mean	Median	TrMean	StDev	SEMean
Carry	1	36	260.19	258.00	257.59	16.76	2.79
	2	36	256.78	257.00	256.78	2.29	0.38

Variable	ModelNo	Min	Max	Q1	Q3
Carry	1	252.00	357.00	256.00	260.00
	2	251.00	262.00	255.00	258.00

Figure 6.10 Descriptive Statistics for *Carry* by Model Number

We may want to further investigate the behavior of the average or mean *Carry* for different time periods within a particular model design. For example, we may want to look at only those values of *Carry* for the M1 ball design for the 8:15 time period. In order to create descriptive statistics for *Carry* for the M1 ball design for the 8:15 time period, you would need to use the **Edit>CopyCells** command sequence explained in Chapter 5. Once you have copied the appropriate values of Carry to another column you can simply use the **Stat>Basic Statistics>Descriptive Statistics** command sequence on that new column.

Section 6.4   Creating Boxplots in Minitab

In the preceding section you learned that you could decide if a particular observation was an outlier by looking at its z-score. A tool which gives you a visual way to see outliers is the boxplot.

As part of the output of the descriptive statistics, Minitab provides you with what is sometimes referred to as the "5 number summary" of the data. These 5 numbers are: the Min(imum), the First Quartile (Q1), the Median (Q2), the Third Quartile (Q3) and the Max(imum). Three of these numbers are used to create a box plot, also called a box and whisker plot.

*Use **Graph>Boxplot** to create a box plot.*

In order to create a box plot in Minitab you must choose Graph from the main menu bar. You have worked with the Graph menu in Chapter 5 and it is redisplayed in Figure 6.11.

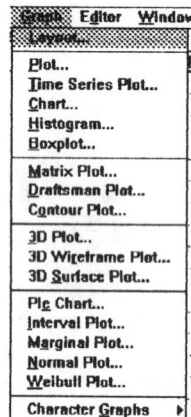

Figure 6.11 The Graph Menu

From this menu select **Boxplot** and Minitab will present you with a **Dialog Box** which is similar to the **Histogram Dialog Box** you worked with in Chapter 5. In the simplest case, if you wish to create a boxplot for one variable without a grouping (or category) variable then you would simply pick a variable from the Variable List to be the **Graph Variable: Y**. However, as we have seen, we are often interested in creating boxplots of a variable such as *Carry* grouped by another variable such as Model number. In this case we need to select a variable from the Variable List to be the **Graph Variable:X.** Notice that for boxplots, Minitab lets us consider using alpha variables for the grouping variable. Thus we can use the column *Model_* (column 2) without the need to convert the model numbers to numeric values. Likewise we can use the variable *Time* as a grouping variable.

The completed **Dialog Box** for creating boxplots of the variable Carry by Model Number is shown in Figure 6.12

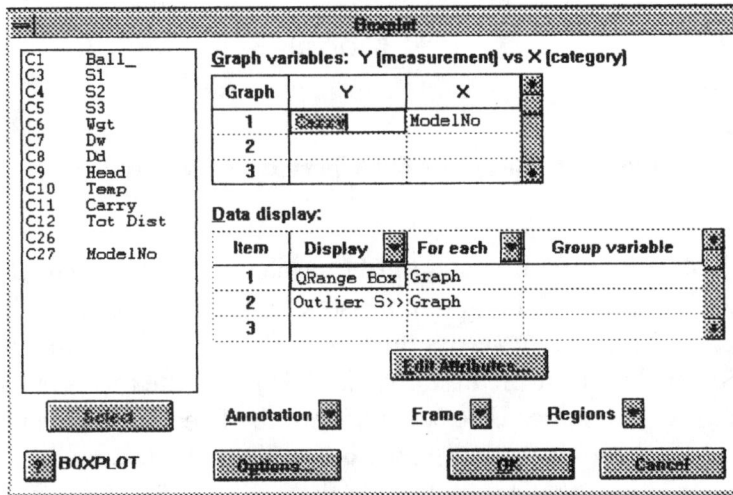

Figure 6.12 Completed Boxplot Dialog Box

Use the **Annotation** pop-up menu to add a title and press the **OK** button to get the boxplots shown in Figure 6.13

Boxplot of Carry by Model Number

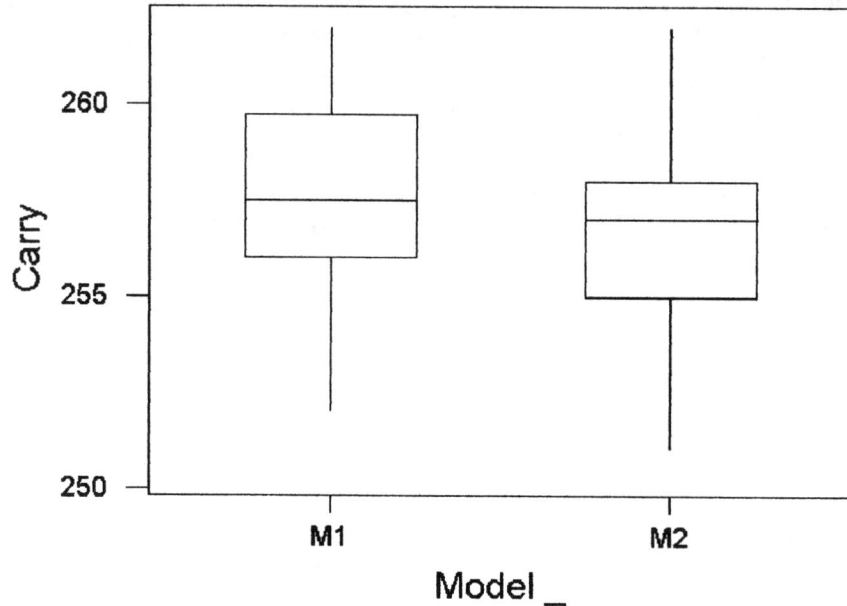

Figure 6.13 Boxplots for *Carry* by *Model Number*

Notice that the first and third quartiles and the median are used to create this graph. A boxplot consists of a box, whiskers, and outliers. A line is drawn across the box at the median. By default, the bottom of the box is at the first quartile (Q1), and the top is at the third quartile (Q3 value. The whiskers are the lines that extend from the top and bottom of the box. they go as far as the lowest and highest data values that are still inside the region defined by the following limits:

Lower Limit: $Q1 + 1.5(Q3-Q1)$
Upper Limit: $Q3 + 1.5(Q3-Q1)$

Outliers are points outside of the lower and upper limits and are plotted by Minitab with asterisks (*). In this case, there are no such outliers. They require further investigation to determine why they are so different from the rest of the data set.

Section 6.5 Investigative Exercises

In this section you will be asked to generate the appropriate descriptive statistics in order to answer the questions about the golf ball design datafile. Remember that the numbers themselves must be interpreted within the context of the specific application. Space has been provided for you to write in the answers to the questions.

1. Complete the following table:

Variable	Mean	Median	Range	Variance	Std.Dev
M1 ball **All times**					
Carry					
Tot Dist					
M2 ball **All times**					
Carry					
Tot Dist					

2. Change the values of *Carry* and *Tot Dist* for Ball# 1 to 357 and 370 respectively. Redo the table with these new values.

Variable	Mean	Median	Range	Variance	Std.Dev
M1 ball **All times**					
Carry					
Tot Dist					
M2 ball **All times**					
Carry					
Tot Dist					

3. Compare the measures of the mean and the median found in exercise 1 with those calculated in exercise 2.

a) What has happened to the mean values as a result of having an outlier in the data set?

b) What has happened to the median values as a result of having an outlier in the data set?

c) What can you conclude about how the mean and the median respond to outliers?

d) How could you alter the calculation of the mean to eliminate this sensitivity to outliers?

4. Study the measures of variability found in exercises 1 and 2.

a) What has happened to the values of the range as a result of having an outlier in the data set?

b) What has happened to the values of the sample variance and standard deviation as a result of having an outlier in the data set?

c) What can you conclude about the usefulness of the range as a measure of variability?

5. Study the relationship between the mean and the median.

a) Find a variable such that the mean and the median are very close (or equal) to each other.   Examine the histogram of this variable. What type of shape does it have?

b) Find a variable such that the mean is larger than the median. Examine the histogram of this variable. What type of shape does it have?

c) Find a variable such that the mean is smaller than the median. Examine the histogram of this variable. What type of shape does it have?

d) Based on these observations what can you conclude about the shape of the distribution based on the relationship between the mean and the median?

5. Compare the mean of the variable *Carry* for ball M1 @8:15AM to the mean of the variable *Carry* for ball M1 @8:45AM. . Notice that their means are very close to each other. Now compare the 2 standard deviations.

a) What do you notice?

b) What does this tell you about the adequacy of only using the mean to describe a data set?

c) Look at a histogram of each of these variables. What can you learn from the graphs about how the two variables differ?

6. Further investigate the variables used in exercise 5.

a) Complete the following table:

Variable	Time	Ball No.	Carry	Distance from its mean	# SD's from it mean
Carry	8:15AM	13	256		
Carry	8:45AM	41	256		
Carry	8:15AM	14	255		
Carry	8:45AM	42	255		
Carry	8:15AM	16	257		
Carry	8:45AM	40	257		

b) What is another name for the values you calculated for the last column?

c) What do you notice about the values in column 5? What does this tell you about the usefulness of simply looking at the distance between an observation and its mean?

7. Is it possible for two variables to have similar variances but vastly different means?

a) Find two such variables in the dataset and create a histogram for each variable.

b) Compare the histograms. If you know that two variables have similar variances but different means, what does that tell you about the behavior of the two variables?

8. Consider the variable *Tot Dist* for the M2 ball for all times.   Look at how well the data for this variable fits the empirical rule. Recall that the empirical rule states that 68% of the data will fall within 1 standard deviation of the mean, 95% of the data will fall within 2 standard deviations of the mean, and 99% of the data will fall within 3 standard deviations of the mean.

a) Display the data as a histogram

b) Complete the following information (from exercise #1):

Count	
Mean	
Standard Deviation	
Mean - 1SD	
Mean+ 1 SD	
Mean - 2SD	
Mean + 2SD	
Mean - 3SD	
Mean + 3SD	

c) Mark the last 6 values on the histogram.

d) How many data values fall within 1 sd of the mean:? _____
Divide this by the number of observations for the variable (N) to get a
%: _____

How many data values fall within 2 sd of the mean? _____
Divide this by the number of observations for the variable(N) to get a
%: _____

How many data values fall within 3 sd of the mean? _____
Divide this by the number of observations for the variable (N) to get a
%: _____

e) How closely do the percentages found in part d) match the empirical
rule? What does this tell you about the general shape of the graph?

f) Consider the value for Ball #2 . Mark it's location on the x-axis of
the histogram. Where does it fall relative to the markings you made in
part c)?

Now find its z-score. What then does the z-score tell you about the
data point?

9a) Can you find a variable in the data set that has a zero (or close to zero) variance?

b) What can you say about the observed values for this variable?

10. Compare the mean for the variable *Carry*, design M1 at 8:45AM to the mean of the variable *Carry*, design M2 at 8:45AM. Notice that they are close in value but not exactly the same.

a) If you observed 12 more values of *Carry* for the M1 design at 8:45AM and calculated the mean of those 12 values, would you get exactly 258.0833? Would the second mean be close to 258.0833? Why or why not?

b) If you observed 12 more values of *Carry* for the M2 design at 8:45AM and calculated the mean of those 12 values, would you get exactly 257.5833? Would you expect it to be close to 257.5833? Why or why not?

c) What does this tell you about the sample means from sample to sample?

d) Now, considering your answer to part c), is it possible that the true underlying population mean of the variable *Carry* for both designs at 8:45AM is the same despite the fact that you have observed 2 slightly different sample means? Why or why not?

Chapter 7   "Credit Problems?"

A Study of The Binomial Distribution

Section 7.1 Overview

In the last three chapters you have learned how to *describe* your data set. When you construct a histogram of your data, what you are looking at is the <u>distribution of the sample</u> data. However, we must remember that it is NOT the sample which we are ultimately interested in but rather the population from which the sample was drawn. Thus, although it is quite helpful to view the distribution of the sample, what we really would like to know is the <u>distribution of the population</u> from which this sample was drawn. You can think of the sample as evidence upon which you wish to draw some *inferences* about the population. In a sense you must be a bit of a detective.

In order to successfully detect the underlying population distribution you must know a little bit about the behavior of some of the distributions which are commonly found. There are two basic classes of distributions: discrete distributions and continuous distributions. This chapter will focus on one of the most commonly used discrete distributions, the **BINOMIAL DISTRIBUTION.** In Chapter 8, you will investigate the most commonly used continuous distribution, the Normal Distribution.

Statistical Objectives: After reading this chapter and doing the exercises a student will:

- Know how to detect a Binomial Variable
- Know how changing the values of n and p affects the shape of the Binomial Distribution
- Be able to makes some statements about the likely value of the parameter p, the probability of success
- Know how to comparing two binomial variables.

Section 7.2 Problem Statement

In recent years the economy has experienced a recession. Banks and other financial institutions have imposed more stringent conditions on companies trying to get loans. This is clearly to protect the banks from making "bad" loans. However, as a result of this conservative policy , many small companies have been forced to close or reduce their size because they can not meet the stiffer requirements needed to get a loan. A medium size New England city has been very concerned about this problem and has tried to find out the specific nature of the credit problems facing small businesses. They have done this by administering a questionnaire every 6 months. The questionnaire consists of many questions which fit the model of the binomial distribution.

The survey was mailed to 1536 companies within a 10 mile radius of this city. A total of 166 usable responses were received. On the basis of your analysis of these 166 responses you will be asked to describe the nature of the credit problem to the Chamber of Commerce.

Section 7.3 Characteristics of the Data Set

FILENAME: CHAP7.MTW
SIZE: COLUMNS   11
                      ROWS   166

The first 7 columns (C1-C7) of the actual datafile are shown in Figure 7.1.

	C1	C2	C3	C4	C5	C6	C7
↓	Number	Size	Employee	Nature	Problem	Understd	Concern
1	1	2	2	1	1	2	1
2	2	1	2	3	1	2	2
3	3	4	3	1	2	1	
4	4	1	1	1	1	2	2
5	5	1	2	1	1	0	0
6	6	3	1	5	2	0	2
7	7	3	3	2	2	1	1

Figure 7.1 First 7 Columns of CHAP7.MTW

The remaining 4 columns (C8-C11) are shown in Figure 7.2.

	C8	C9	C10	C11	C12	C13	C14	
↓	Call	Loan	Collater	Access				
1	2	2	0	2				
2	0	2	2	1				
3	2	2	2	0				
4	2	2	2	2				
5	0	0	0	0				
6	2	2	2	1				
7	2	2	2	1				

Figure 7.2 Last 4 Columns of CHAP7.MTW

Notes on the dataset:

For all variables the code of zero (0) was used when the question was not answered by the respondent.

1. The variable *Number* simply numbers the respondents from 1 to 166.

2. The variable *Size* refers to the annual sales of the company and is coded as follows:
1	Under $1 million
2	$1- 5 million
3	$5- 10 million
4	$11- 20 million
5	Over $20 million

3. The variable *Employee* refers to the number of employees which the company currently employs. This variable was coded as follows:
1	0-5 employees
2	6-10 employees
3	11-50 employees
4	51-150 employees
5	151-250 employees
6	Over 250 employees

4.   The variable *Nature* refers to the nature of the business and is coded as follows:

1	Manufacturing
2	Retail
3	Service
4	Real Estate
5	Other

The coding scheme for the remaining variables is as follows:

1	Yes
2	No

5.   The variable *Problem* contains the respondents answer to the question: "Are you experiencing recession related problems?"

6.   The variable *Understd* contains the respondents answer to the comment: "Bank understands my business".

7.   The variable *Concern* contains the respondents answer to the comment: "Concerned my note may be recalled".

8.   The variable *Call* contains the respondents answer to the comment: "Bank is planning to call my loan".

9.   The variable *Loan* contains the respondents answer to the comment: "Bank has called my loan".

10.   The variable *Collater* contains the respondents answer to the comment: "Bank has demanded more collateral".

11.   The variable *Access* contains the respondents answer to the comment: "Access to credit is effecting my business".

*Use **File>Open Worksheet** to read the file.*

Read in the file named CHAP7.MTW using the commands described in Chapter 3. Remember to create a working version of the datafile with a slightly different name.

Section 7.4 Detecting a Binomial Variable

In order to analyze this dataset, you must first look at whether or not any of these variables fit the model for a Binomial Random

Variable. Recall that a Binomial Random Variable has the following properties:

- The experiment consists of n independent trials;
- Each trial results in one of two possible outcomes; a successful outcome and a failure outcome;
- The probability of a successful outcome is the same from trial to trial and is called p;
- The probability of a failure outcome is the same from trial to trial and is called q= 1-p.

Consider the variable *Problem* as described above. Do you think this variable fits the model for a Binomial Random Variable? Let's have a look!

The "experiment" consists of 166 independent responses to the question: "Are you experiencing recession related problems?". They can be considered independent unless we have reason to believe that the respondents influenced each other in answering the questionnaire. There is no reason to believe this. So each response to this question is one "trial" and we have met the first criteria.

The possible responses to this question were either Yes or No. If we consider the response "Yes" as the successful outcome and the response "No" as the failure outcome, then we have met the second criteria.

The probability of a successful outcome is then the probability that the response is "Yes" from any random respondent. This probability is unknown. In order to satisfy the third and fourth criteria, you must see if the value of p (and therefore q) is the same from trial to trial.   If we did know p, it would be equal to the percentage of the 1536 members of the population who are experiencing recession related problems. Pretend for a moment that you do know the value of p. Then if you randomly selected a respondent from the population, the chance that it would be experiencing recession related problems would be p. If you made another random selection, would the chance that it was experience recession related problems still be p? What do you think?

Unless you sampled with replacement, that is you allowed the first respondent to be picked again, the answer would have to be no the value of p would not be the same as on the first pick. Thus we are in violation of criteria 3 and 4. In most actual situations you will not sample with replacement as it does not make much sense to do this. So technically speaking you almost never have a true binomial random variable. In this case we can just skip the rest of this chapter, right? No! This is the first instance in this workbook where we must use our understanding of statistics to decide if we can "bend the rules a bit and still be OK". This will not be the last time we encounter this. In actual data situations you almost never meet all of the criteria or assumptions of the theory. Thus, it is vitally important that you understand the theory well enough to decide what rules you can and cannot bend!

In this case we must decide if sampling without replacement yields a situation where the value of p, although changing from trial to trial, is not changing by much. Suppose for example that there are 231 out of 1536 companies experiencing recession related problems. Then on our first selection, the chance of picking someone who is experiencing a recession related problem is 231/1536 = .15039. On our second selection, the probability of selecting someone who is experiencing a recession related problem is either 231/1535= .15048 or 230/1535=.14983. As you can see, the value of p is changing but not by much.   So we accept the slight deviation from the exact model and use the binomial distribution to model the variable *Problem*.

Exercise 1. What other variables in this data set can be considered binomial variables? For each variable, identify the successful outcome, the failure outcome and the value for n.

Now that we have decided to use the binomial distribution, we must decide how to use it since we do not know the value of p. In fact, your job is to make some statements about the likely value of p for each of the binomial variables. In order to do this we need to have a good understanding of what the binomial distribution looks like and how it behaves. We will use Minitab to save us the tedious calculations.

Section 7.5 The Binomial Distribution in Minitab

Section 7.5.1 Creating Binomial Probabilities

The first step in generating a binomial distribution in Minitab is to create a column of numbers which gives all of the possible values of binomial random variable. Remember that the binomial random variable, X, is defined to be the total number of successes in n trials. Thus X can be 0, 1, 2, 3, and so on all the way up to n, where n refers to the total number of trials in the experiment. This is the sample size. Thus, in order to tabulate the entire distribution you would start at 0 and end at n. After all you can't have less than zero successes or more than n successes in a sample of size n! If we continue to examine the variable *Problem* in this dataset, then n=166.

So the first step is to create a column of numbers starting with 0 and going up to 166. To do this in Minitab select **Calc** from the main menu bar. The **Calc** menu is shown in Figure 7.3.

First, create a column of integers from 0 to n.

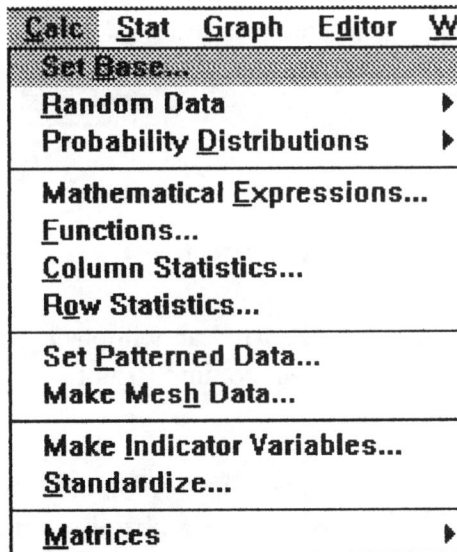

```
Calc   Stat   Graph   Editor   W
Set Base...
Random Data              ▶
Probability Distributions ▶
Mathematical Expressions...
Functions...
Column Statistics...
Row Statistics...
Set Patterned Data...
Make Mesh Data...
Make Indicator Variables...
Standardize...
Matrices                 ▶
```

Figure 7.3 Calc Menu

From this menu select **Set Patterned Data** and you will see a dialog box. By clicking on the **Patterned sequence** button and specifying the **Start at** and **End at** values you can tell Minitab to create a column beginning with the number 0 and ending with the value of 166. You must also specify a column location for Minitab to place the results. In this case, column c12 is the next empty column so that has been indicated in the **Store result in column** portion of the dialog box. The completed Set Patterned Data dialog box is shown in Figure 7.4

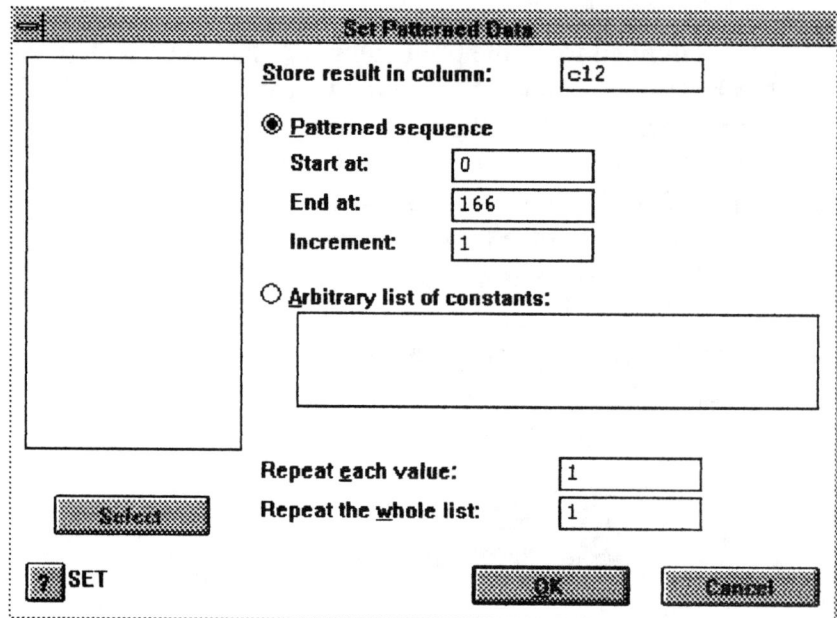

Figure 7.4 The Set Patterned Data Dialog Box

When you click on **OK**, Minitab will place the values of 0,1,2,3 and so forth up to 166 in column C12 of the worksheet. Label this column with the name X.

*Use **Calc>Probability Distributions** to calculate probabilities*

The second step in generating a binomial distribution in Minitab is to have Minitab calculate a probability value for every possible value of X, the number of successes. This means we want to create a column of probabilities which correspond to the possible values of X now stored in column C12 in our worksheet. Again select **Calc** from the main menu bar and you will see the **Calc** menu shown in Figure 7.3. From this menu select **Probability Distributions** and you will see the pop-up menu shown in Figure 7.5

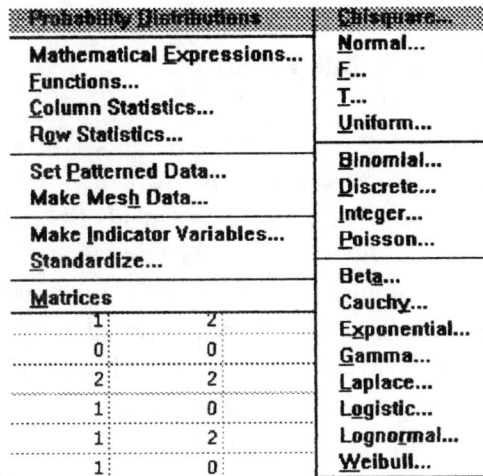

Figure 7.5 Probability Distribution Pop-up Menu

From this menu choose **Binomial**. Minitab will then give you the Binomial Distribution Dialog Box shown in Figure 7.6.

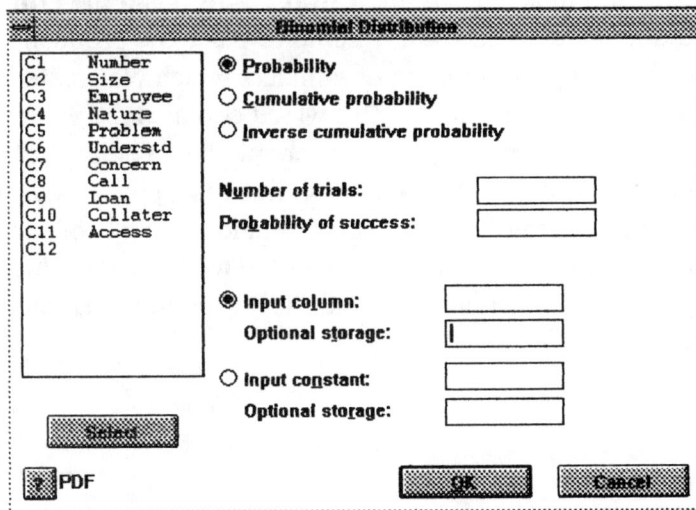

Figure 7.6 Binomial Distribution Dialog Box

As you can see in Figure 7.6, you have several choices for the type of probability you would like Minitab to calculate. In this chapter we will concentrate on using the **Probability** and **Cumulative**

probability distributions. In Chapter 8 you will see the use of the **Inverse cumulative probability** feature.

Click on the Probability button to calculate probabilities.

If you want Minitab to calculate the probability of observing *exactly* 0 successes, 1 success, and so forth then you should click on the **Probability** button in this dialog box (this is the default). For our study of the variable *Problem*, this means Minitab will calculate for us the probability that we would observe 0 "Yes" responses to the question "Are you experiencing recession related problems?" out of the 166 responses. Then it will calculate the chance of observing 1 "Yes" response to this question out of 166 and then the probability of observing 2 "Yes" responses and so forth all the way up to the probability of observing 166 "Yes" responses out of 166. In order to do so you must also tell Minitab the values for **Number of trials**, n and **Probability of success**, p.   In our case n is 166 and we need to consider different values of p. In data analysis p will be unknown. For now, we will look at the behavior of the Binomial Distribution for various values of p. For now use p=.3.

Lastly you must specify the **Input column** and the **Optional storage** column where you want Minitab to place the probabilities. The **Input column** is the column of numbers which we created in the first step. In our case it is column C12 which contains the values from 0 to 166. The **Optional storage** column should be an empty column, preferably next to the **Input column**. In this case, specify C13 as the location for the probabilities. Click on **OK** and you will see the probabilities stored in the worksheet in column C13. Label this column P(X). A portion of the columns X and P(X) (columns C12 and C13) are shown in Figure 7.7.

40	0.0171724
41	0.0226174
42	0.0288487
43	0.0356535
44	0.0427148
45	0.0496305
46	0.0559499
47	0.0612218
48	0.0650482
49	0.0671343
50	0.0673261
51	0.0656288
52	0.0622031
53	0.0573409
54	0.0514248
55	0.0448798

**Figure 7.7 A portion of the Binomial
Distribution for n=166 and p=.3**

Column C12, which we labeled X, gives the values of the random variable. Since we are considering the variable *Problem*, this column corresponds to various possible values for the number of respondents who answered Yes (the successful outcome) to the question: "Are you experiencing recession related problems?". Clearly anywhere from 0 to 166 might have responded "Yes" to this question.

Column C13, which we labeled P(x), gives the probability corresponding to the value of X shown in column C12. For example, if you look at the row where X equals 40 then you will see the probability is .017172. This is the probability you would get if you plugged the values x=40, n = 166 and p=.3 into the formula for the binomial distribution. As you examine column c13 you will see lots of zeros and small probabilities at the top and bottom of the column. This is because it is unlikely that you would observe a relatively few number of "yes" answers out of 166 and also it is also unlikely that you would observe mostly "yes" responses when p=.3. In fact, you have probably noticed that most of the non-zero probability lies in the range $20<X<60$. You can tell this by observing the teeny probabilities in the second column for values of X outside this range.

Exercise 2. What is the probability of observing exactly 45 "Yes" responses for the variable *Problem* if p=.3?

Section 7.5.2 Creating Cumulative Binomial Probabilities

In the previous section we learned how to calculate Binomial probabilities using Minitab. We noticed that in the Binomial Distribution dialog box (shown in Figure 7.6) that we could also easily obtain cumulative probabilities. Let us look at those now.

Click on the Cumulative probability button to calculate cumulative probabilities.

Choose **Calc>Probability Distributions>Binomial** and this time click on the button next to the words **Cumulative probability** in the dialog box. Specify column C14 as the **Optional storage** column and leave everything else the same as before. Click on **OK** and you will see the cumulative probabilities in column C14 of the worksheet. Label this column C(x). A portion of the columns X, P(x) and C(x) are shown in Figure 7.8 showing the values of X from 40 to 55, the corresponding probabilities and the cumulative probabilities.

X	P(x)	C(x)
40	0.0171724	0.05543
41	0.0226174	0.07805
42	0.0288487	0.10690
43	0.0356535	0.14255
44	0.0427148	0.18527
45	0.0496305	0.23490
46	0.0559499	0.29085
47	0.0612218	0.35207
48	0.0650482	0.41712
49	0.0671343	0.48425
50	0.0673261	0.55158
51	0.0656288	0.61721
52	0.0622031	0.67941
53	0.0573409	0.73675
54	0.0514248	0.78818
55	0.0448798	0.83306

Figure 7.8 A portion of the Binomial probabilities and the Cumulative probabilities for n= 166 and p=.3

You should notice that the values in the cumulative probabilities column are increasing. The third column, labeled C(x), gives the cumulative probability distribution function. For the value of X=55 the cumulative probability function is $P(X < =55)$ and is given in the table as .83306. This means there is about an 83% chance that you would see at most 55 "Yes" responses if the value of p was equal to .3.

Exercise 3. Using the information shown in Figure 7.8, find the probability of observing 43 "Yes" responses if p is actually .3.

Exercise 4. Find the probability of observing 50 or less "Yes" responses, if p is .3.

Exercise 5. Find the probability of observing more than 50 "Yes" responses, if p is .3.

Section 7.6 Plotting the Binomial Distribution in Minitab

What does the graph of this binomial distribution look like? To display a graph of the distribution shown in the column labeled P(x), column c13 we will a bar chart. You learned the basics of creating bar charts in Chapter 4.

Select **Graph** from the menu bar and choose **Chart**. The Chart Dialog Box will appear. In the **Graph variables** portion of the dialog box you should specify that Y variable is found in column C13, the Binomial probabilities and the X variable is found in column C12, labeled X. Next you should specify that the function you wish to perform. Click on the arrow next to the word Function and select Sum from the pop-up menu which appears.

In the Data display area of the Dialog Box the default display of a bar chart for each graph can be left as it appears. Add titles and information about how the axis are to appear by using the Annotation and Frame pop-up menus as explained in Chapter 4. The completed dialog box is shown in Figure 7.9.

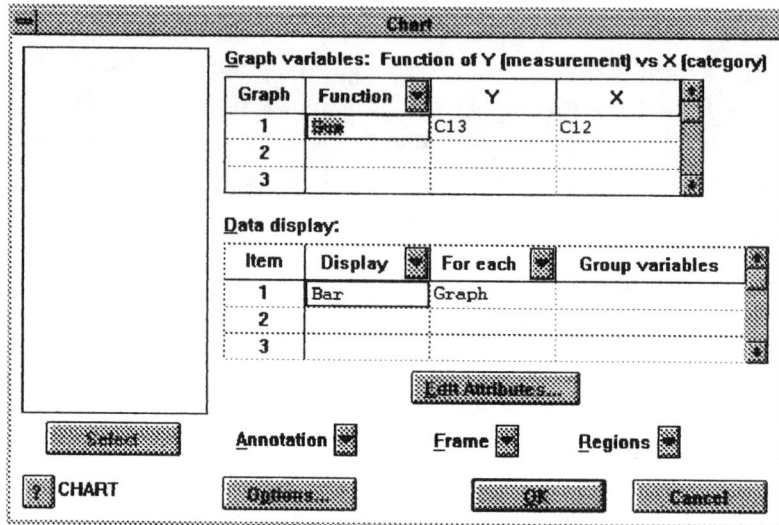

Figure 7.9 Completed Chart Dialog Box

The completed title, axis and tick dialog boxes are shown in Figure 7.10.

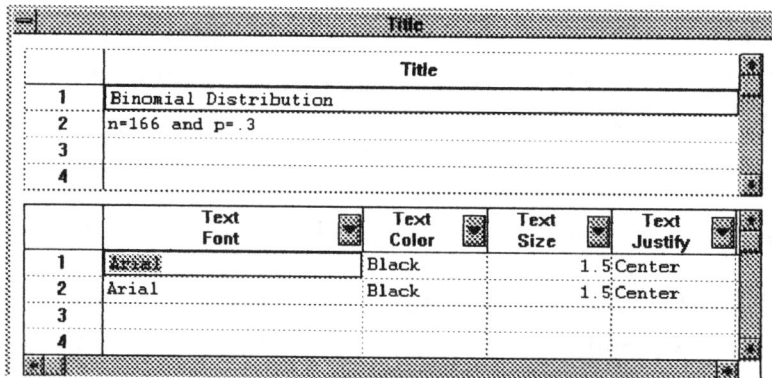

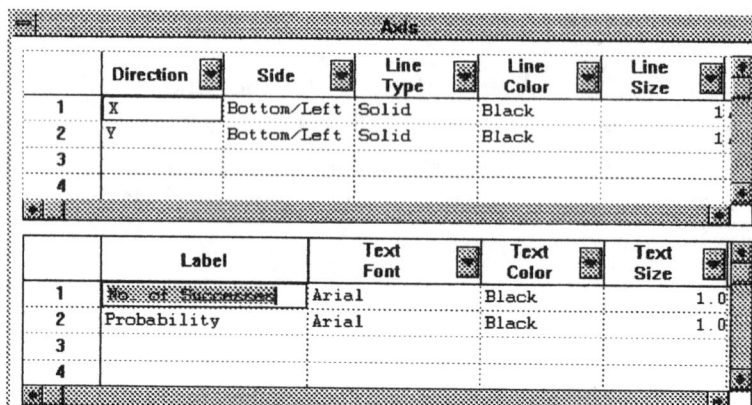

	Direction	Side	Positions	Number of Major	Number of Minor	
1	X	Bottom/Left	Auto	auto	auto	
2	Y	Bottom/Left	Auto	Auto	Auto	
3						
4						

	Text Color	Text Size	Text Angle	Horizontal Offset	Vertical Offset	
1	Black	0.5	90	0.0	−0.05	
2	Black	1.0	0	Auto	Auto	
3						
4						

Figure 7.10 Completed Title, Axis and Tick Dialog Boxes

When you have completed all the dialog boxes then click **OK** in the main Chart dialog box and you will see the graph shown in Figure 7.11

Binomial Distribution
n=166 and p=.3

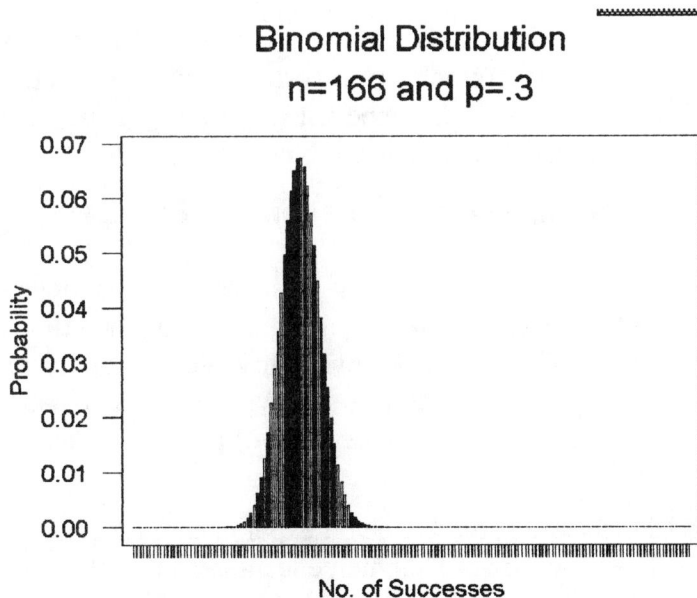

Figure 7.11 Binomial Distribution for p=.3 and n=166

It is important for you to have an understanding of what the binomial distribution looks like However, it is equally as important to realize that if you construct a bar chart of the responses to any of the binomial variables in this data set it will NOT look like the graph shown in Figure 7.11, nor should it. They are completely different graphs.

Exercise 6. Construct a bar chart of the responses in the variable *Problem.*

Notice that the bar chart simply gives a visual display of how many respondents answered Yes, No or gave No Response to the question: "Are you experiencing recession related problems?". This graph does not have as its X axis the number of successes in 166 responses as the graph in Figure 7.11 does. They are NOT comparable graphs. The graph of the binomial distribution shows how likely it is that you would get x successes if p =.3. The graph you made in Exercise 6 shows that there were 25 successes (Yes responses) for this survey. Thus x = 25 for this survey. It corresponds to one of the 166 different possible values of x.

This is not to say that you should not create a bar chart of the responses to the survey. Rather it is important to understand that when you do so you should NOT expect it to have the shape of a binomial distribution even though it is a binomial variable.

Section 7.7 Estimating the Probability of Success, p

Much of our discussion thus far in the chapter has focused on how to create a binomial distribution table and graph when the value of p, the probability of success is known. However, if you knew the value of p you would not have done a survey! This means that we must somehow be able to estimate the value of p.

Consider the question "Are you experiencing recession related problems?". The Chamber of Commerce is interested in the percentage of respondents who answered "Yes" to this question. Thus, it makes sense to make the "successful outcome" the "Yes" response. We could estimate p by taking the number of Yes responses and dividing by the total number of respondents.

Why do you think this might be misleading? Suppose of the 166 responses, there were 25 Yes's, 45 No's and 96 respondents who did not answer the question. Using the method just suggested would yield an estimated p value of 25/166 = .15. This is misleading because of the 70 responses to this question 25 of them were Yes responses. Thus, a more accurate reflection of the survey result would be to estimate p as 25/70= .36. This is considerably different and gives a much different picture of the financial situation facing the companies in this New England city. Clearly you should do some investigation as to why there was such a high non-response rate to this question.

In order to estimate the value of p using Minitab you need to have Minitab count for you how many Yes responses (coded 1) there were, how many No responses (coded 2), and how many respondents did not answer the question (coded 0). From the menu bar click on **Stat**. The pull-down menu shown in Figure 7.12 will appear.

Stat	Graph	Editor
✓Fit Intercept		
Basic Statistics		▶
Regression		▶
ANOVA		▶
DOE		▶
Control Charts		▶
SPC		▶
Multivariate		▶
Time Series		▶
Tables		▶
Nonparametrics		▶
EDA		▶

Figure 7.12 The Stat Menu

From this menu choose **Tables** and the pop-up menu shown in Figure 7.13 will appear.

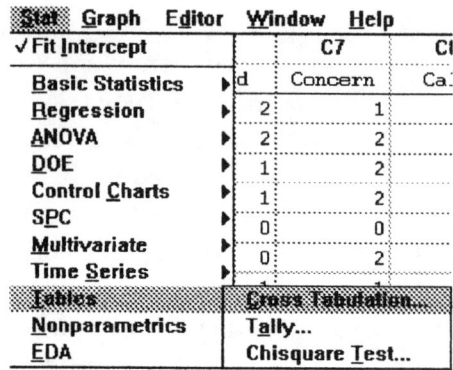

Figure 7.13 Tables Pop-up Menu

*Use
Stat>Tables>Tally to
find raw counts of each
of the possible values
of the variable.*

From this menu choose **Tally** and a dialog box will be shown.
Complete the dialog box as shown in Figure 7.14 to tell Minitab to
find raw counts of each of the possible values of the variable *Problem*.

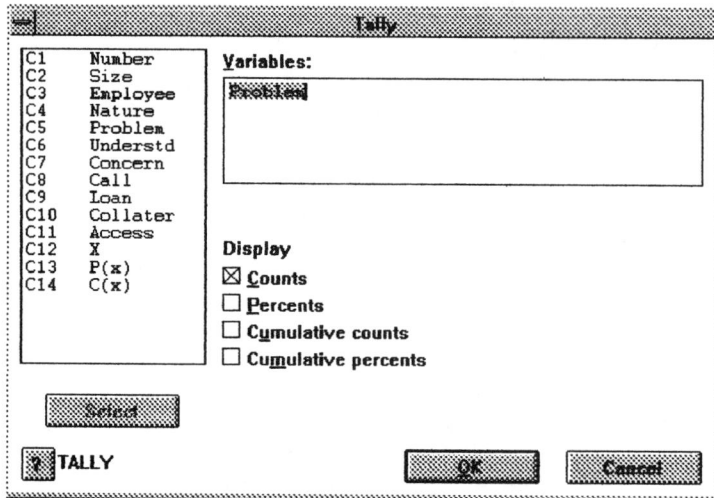

Figure 7.14 Completed Tally Dialog Box

Click on **OK** and Minitab will display the counts in the Session
window and it will look like that which is shown in Figure 7.15.

Summary Statistics for Discrete Variables

Problem	Count
0	16
1	25
2	125
N=	166

Figure 7.15 Counts for the variable *Problem*

Using the information shown in Figure 7.15 you can estimate p=probability of a Yes response = 25/(166-16)= 25/150=.167.

When you tell the Chamber of Commerce that 16.7% of the respondents are experiencing recession related problems they might want to know if small businesses are being "hit" harder than larger businesses.   In order to answer this question you find what are know as cross tabulations. This is simply a table with the various possibilities for the variable Problem shown as the rows and the various possibilities for the variable Size shown as the columns.

To do this in Minitab choose **Stat**>**Tables**>**Cross Tabulation** (see Figure 7.13). A Cross Tabulation dialog box will appear and you must simply specify the two **Classification variables**. In our case we select *Problem* and *Size* and click on the button next to **Counts**. the completed dialog box is shown in Figure 7.16.

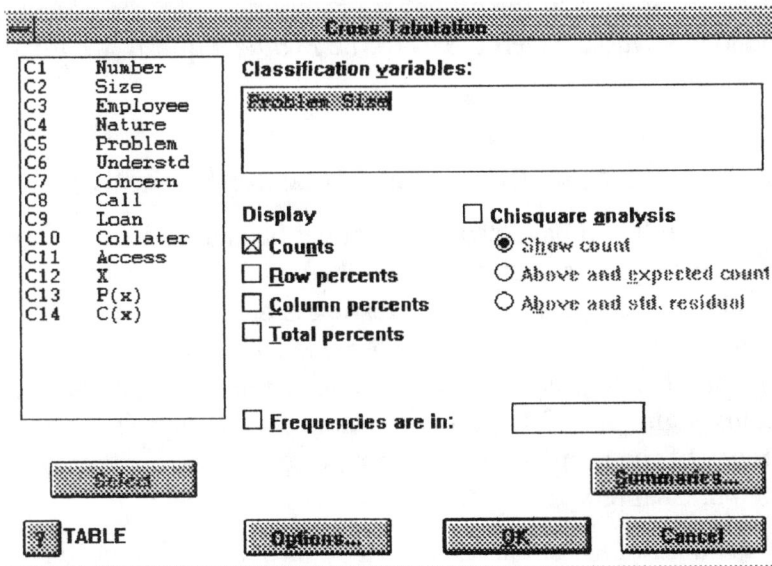

Figure 7.16 Completed Cross Tabulation Dialog Box

Click on the **OK** button and you will see the table shown in Figure 7.17 in the Session window.

```
ROWS: Problem     COLUMNS: Size

              0       1       2       3       4       5     ALL

      0       0       7       0       4       2       3      16
      1       0       6       7       5       3       4      25
      2       1      47      14      33      16      14     125
    ALL       1      60      21      42      21      21     166

    CELL CONTENTS --
                  COUNT
```

Figure 7.17 Cross Tabulation of *Problem* by *Size*

From the table shown in Figure 7.17 we can see that there are 53 respondents Under $1 million in *Size* (*Size* =1)and who answered either Yes (coded 1) or No(coded 2) to the *Problem* question. Of these 53 companies, 6 or 11% are experiencing recession related problems.

Exercise 7. For each of the other *Size* categories, find the percentage of respondents who answered Yes to the *Problem* question. What can you conclude?

Section 7.8 The Beginnings of Hypothesis Testing

Although you have probably not yet been introduced to hypothesis testing, it is useful to study the binomial distribution with this technique in mind. Hypothesis testing is one of the most commonly used tools of statistical inference and it is helpful to start viewing data this way as early in the game as possible. For this reason, a preliminary glimpse of hypothesis testing is given in this section without any of the terminology or formal procedures.   These will be discussed in Chapters 10, 11 and 12.

In addition to estimating the value of p, it is often necessary to answer the following type of question: *Is it reasonable to believe that the true value of p is .50?* If you had an estimate of $\bar{p}$= .53 from the survey data then it would be easy to say yes to this question. Now suppose that $\bar{p}$ =.20. Then the answer to this question is easily no. However, suppose $\bar{p}$ =.57. Now what is your answer to this question? It gets a little bit harder to decide if .50 is the true proportion of successes in the population given that you have observed 57% successes in the sample. Remember that the sample is just a piece of the population. It is possible that the proportion of successes in the population is only 50% but in your sample it was 57%. Having established that it is possible, the right question is "how *likely* is it that I would observe a $\bar{p}$ of .57 IF the true proportion is .50.

Minitab seems to have trouble with p=.5

In order to answer this question, you need to use Minitab to calculate the likelihood or probability of the observed sample if p really is .40.   Again consider the variable *Problem* . Using the information that you found in the previous section, you know that there were 150 non-zero responses to this question and that 25 of these were Yes responses. You need to calculate the probability of observing 25 or fewer successes if the true value of p is .40. To do this you simply generate a column of cumulative binomial probabilities for p=.40 and n=150. A portion of this column is shown in Figure 7.18.

25	0.00000
26	0.00000
27	0.00000
28	0.00000
29	0.00000
30	0.00000
31	0.00000
32	0.00000
33	0.00000
34	0.00001
35	0.00001
36	0.00003
37	0.00006
38	0.00012
39	0.00023
40	0.00044

**Figure 7.18 Cumulative Binomial Probabilities
for n=150, p=.40**

The probability of observing 25 or fewer successes is close to zero. This small probability tells you that it is quite unlikely that the true value of p is .40.

Exercise 8. For the variable *Problem*, find the probability of the observed sample IF p = .10, .20, .30. Summarize your results in the table shown below.

True p= Probability of Success	Probability of Observed Sample
.10	
.20	
.30	
.40	.00000

        As you can see the question which was posed at the beginning of this section can be asked for any value of p. It should be noted that many times investigators are in fact interested in knowing if it is likely that the true value of p is .50. This is simply because a p value of .50 implies that the population is evenly split between Yes's and No's. In many cases this indicates that there is nothing of interest happening as a purely random situation would yield 50% successes and 50% failures. Thus an investigation of p=.50 is often done to see if there is anything other than random variation.

        In this section you have seen the beginnings of the tool called hypothesis testing. This tool will be looked at in much greater detail in Chapters 10, 11, and 12.

Section 7.9 Investigative Exercises

1. Investigate the behavior of the binomial distribution for p=.3 by varying the value of n.

a) Generate and plot the binomial distribution for p=.3 and n=10.

b) Repeat part a) for p=.3 but let n= 15, 20, 25, 30, 50, 100.

c) Describe the shape of the distribution as n increases.

2. Investigate the behavior of the binomial distribution for n=30 and p=.05, .20, .35, .50, .65, .80, .95

a) In each case, generate the probability distribution and then look at the graph.

b) Describe what happens to the shape of the distribution as p varies from a small value such as .05 to a larger value such as .95.

c) Examine the graphs for p=.20 and p=.80. What do you notice? Now look at the graphs for p=.05 and p=.95. Do you see the same type of relationship?   Make a general statement which reflects what you have noticed.

d) What is special about the graph for p=.5?   Why should you have expected it to look this way?

3. Continue the investigation of the variable *Problem* which was started in the chapter. Are there any differences among the percentage of respondents experiencing recession related problems for the various industries?

4. Summarize what you have found out about the variable *Problem*. This should include an estimate of p for the entire population, estimates of p for the population broken down by size and by industry as well as some discussion of any differences you find.

5. Investigate the variable *Understands*. Your analysis should follow the same format that was used for the variable *Problem*.

6. Investigate the variable *Concerned*.

7. Investigate the variable *Call* .

8. Is it likely that the true proportion of businesses who feel the bank does not understand their business is .20? .30? .40? .50? .60? .70?

9. Is it likely that the true percentage of companies in this New England City who are concerned that their note may be recalled is .50?

10. What recommendations would you make to the Chamber of Commerce.

Chapter 8   "Tissue Strength"
A Study of the Normal Distribution

Section 8.1 Overview

Statistical Objectives: After reading this chapter and doing the exercises a student will:
- Know the meaning of a z-value from a standard normal distribution
- Know how to calculate left and right tail probabilities from a normal distribution.
- Know how changing the mean and standard deviation of a normal distribution affects the left and right tail probabilities.
- Know how changing the mean and standard deviation of a normal distribution affects the shape of the distribution.
- Know how to compare real data to a theoretical normal distribution.

Section 8.2 Problem Statement

Large companies solicit consumer complaints to try to correct problems in the manufacturing process that contribute to the number of complaints. Customers feel that the company is listening to them, and the company can try to correct problems before they lose customers to competitors.

In Chapter 4 we learned about some of the customer complaints that a company that manufactures tissues can get. One of the categories of complaints that made up a large percentage of the total was Dispensing. In that category, Sheets Tear on Removal was a significant factor.

The management of the tissue company has decided to address this problem. They know that tensile strength is the factor that determines when a tissue will tear, and have decided that to solve the problem they will have to investigate the tensile strength of the tissues.

As part of the Quality Control program at the company, facial tissue has certain product specifications, that is, criteria that must be met, in order for the product to be acceptable to consumers. One of the characteristics that is specified is tensile strength.

The group has decided to look at the current levels of tissue strength. They know the target values for the process, and the parameters that should be met, and have decided to check to see if the process is meeting the current specifications. If it is not, then changes will need to be made to see that is does. If it is, then perhaps the process specifications will need to be changed. They will collect data on two variables:

Machine Direction (MD) Strength: This is the strength in the direction that the machine pulls on the tissue during manufacture. It has to be high enough that the tissues do not break, causing machine down time.

Cross Direction (CD) Strength: This is the strength is the direction that the tissue is pulled out of the box. It is the variable that would control Sheets Tear.

One of the other variables that needs to be considered is some measure of total strength, which is related to both the dispensing defects and another important tissue parameter, softness. Often strength and softness are tradeoffs. Using both MD and CD Strength, they can calculate the **Geometric Mean Tensile (GMT),** which is equal to the square root of the product of the two variables. This variable is also subject to process specifications.

Samples were taken from tissue produced on a single tissue machine. The samples were taken over three different days and the results were recorded.

Section 8.3 Characteristics of the Data Set

FILENAME:	CHAP8.MTW	
SIZE:	COLUMN	5
	ROWS	200

The first seven lines of the datafile is shown in Figure 8.1.

	C1	C2	C3	C4	C5	C6	C7
↓	DAY	MDSTRENG	CDSTRENG				
1	1	1006	422				
2	1	994	448				
3	1	1032	423				
4	1	875	435				
5	1	1043	445				
6	1	962	464				
7	1	973	472				

Figure 8.1 The Tissue Strength Datafile

Notes on the Datafile:

1. The variable *Day* keeps track of the day on which the sample was taken and goes from 1 to 3.

2. The variable *MDStrength* measures Machine-directional Strength and is measured in lb./ream.

3. The variable *CDStrength* measures Cross-directional Strength and is measured in lb./ream.

Load the datafile into Minitab and save it again under a different name.

Section 8.4 Normal Distribution in Minitab

In order to determine whether the tissue being produced meets the specifications of the manufacturing process is necessary to see what the theoretical distribution of tissue strength looks like. Very often, measurement data is assumed to be normally distributed. This is very often true not only because of the nature of the data itself, but because of the nature of physical measurement. If a single item is measured for some characteristic such as length, the measurements will vary due to *measurement error*. Most measurements will center on the true value, but some will be higher and some lower. The further a measurement gets from the true value, the less likely it is to occur. This is in fact characteristic of the normal distribution.

According to the specifications, the MD Strength is supposed to be normally distributed with a mean of 1000 and a standard deviation of 50 lb./ream. The CD Strength should be normally distributed with a mean of 400 and a standard deviation of 25 lb./ream.

We can use Minitab to create a table of values from a normal probability distribution with a given mean and standard deviation (general normal). Usually when you want to find probabilities for normally distributed random variables, you have to convert the value of X to a z-value and then look up the answer on a standard normal table. Minitab provides a table of probabilities in terms of the original random variable, X. In order to do this you must first create a set of X values for which to evaluate the probabilities. From the main menu bar select **Calc** > **Set Patterned Data**. The dialog box is shown in Figure 8.2.

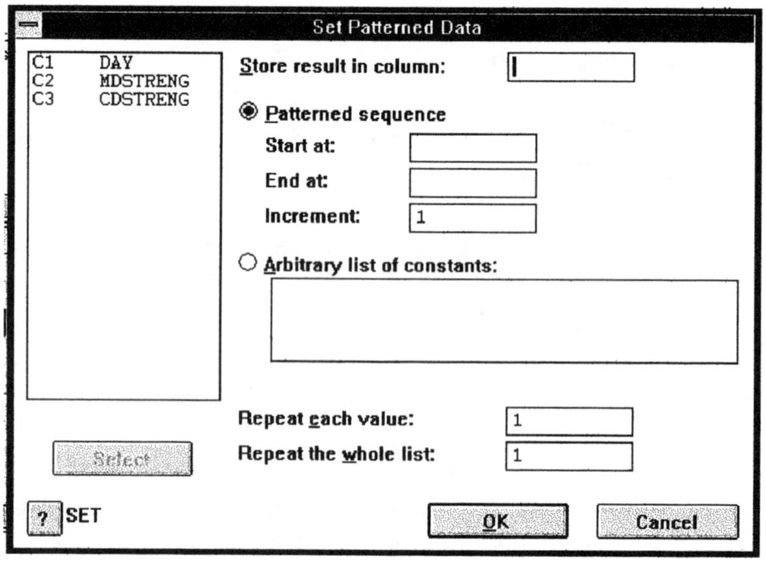

Figure 8.2 The Set Patterned Data Dialog Box

Position the cursor in the text box labeled **Store results in column:** and enter the location of an empty column in the worksheet. In addition to a location you need to supply Minitab with a START, END and INCREMENT for the data that is to be created.

When you used this function for the Binomial Distribution in Chapter 7 the choices for the beginning, ending and step values for the table were obvious. To decide on the same values for a normal density function requires a bit more thinking.

For a normal distribution, 99.73% of the area under the curve lies within 3 standard deviations of the mean. Just to make sure that you have all of the area included in the table start your table 4 standard deviations below the mean and end it 4 standard deviations above the mean. To get these values you will have to do the following calculations:

START = MEAN - 4 * (STANDARD DEVIATION) (Eq. 8.4.1)

END   = MEAN + 4 * (STANDARD DEVIATION)   (Eq. 8.4.2)

When you use the values for the MD Strength you should get START = 800 and STOP = 1200. (Make *sure* you get these values. You will have to do similar calculations later in the chapter.) Position the cursor in the text box for **Start at:** and enter 800. In the text box for **End at:** enter 1200.

The INCREMENT value controls the number of entries in the table. For this exercise we will choose an increment of 25. This means that Minitab will create a column of data that starts at 800 and goes to 1200 in increments of 25 (800,825, ..., 1175, 1200). Enter 25 in the appropriate text box. Click on **OK** and you will see that a new column of data is added to the worksheet at the location you specified.

Now you are ready to create the table of normal probabilities that correspond to this data. From the main menu bar select **Calc > Probability Distributions > Normal...** The dialog box is shown in Figure 8.3.

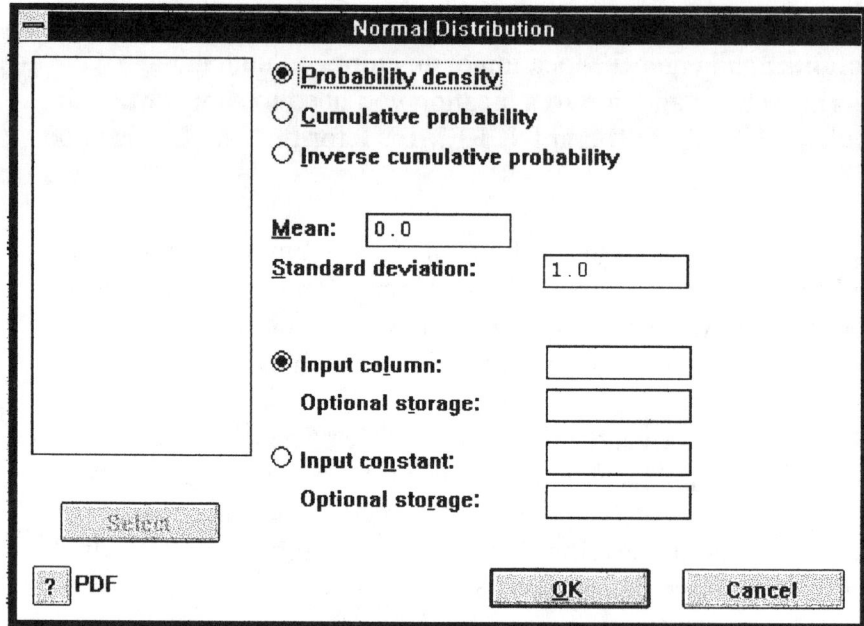

Figure 8.3 Normal Distribution Dialog Box

Remember, for a continuous random variable the probability distribution is referred to as a density function

Minitab gives you the option of generating values of the **Probability density,** the **Cumulative probability** and the **Inverse cumulative probability**. To create a table of probability densities for a normal distribution, click on the option button for **Probability density**. To define a normal distribution you need to specify a mean, μ, and a standard deviation, σ. Enter the values for the variable *MDStrength,* 1000 and 50, in the appropriate text boxes.

You can designate a column in the worksheet using Cn notation or by naming it. If the named variable does not already exist, Minitab will choose the next empty column in the worksheet.

Since the data values are located in a column, click on the option button for **Input column**. In the text box enter the location of the data values you created. Minitab will either provide the probabilities in the **Session** window or put them in a column in the worksheet. Since you will want to use these values later on, position the cursor in the text box labeled **Optional storage:** and enter a location for the probabilities. Click on **OK** and the column will be added to the worksheet.

Exercise 1. Create another column in the worksheet for the values of the **cumulative probability** distribution.

When you have done this, the worksheet should look like the one in Figure 8.4.

	C1	C2	C3	C4	C5	C6	C7
↓	DAY	MDSTRENG	CDSTRENG	X	ProbDist	CumDist	
1	1	1006	422	800	0.0000027	0.000032	
2	1	994	448	825	0.0000175	0.000233	
3	1	1032	423	850	0.0000886	0.001350	
4	1	875	435	875	0.0003506	0.006210	
5	1	1043	445	900	0.0010798	0.022750	
6	1	962	464	925	0.0025904	0.066807	
7	1	973	472	950	0.0048394	0.158655	

Figure 8.4 Worksheet After Adding Normal Probabilities

Note: If your normal table gives the area under the normal curve from 0 to z then the value you find there will be 0.5 less than the value in the CumDist column.

Look at the last column you generated (cumulative probability) carefully. These values are the probabilities that a tissue chosen at random will have an MD Strength that **is less than or equal to** the value in the first column of the table. Check the value for *CumDist* for an MD Strength of 1000. It should be .5. This will correspond to the value in a Standard Normal Table for z = 0. Check the value of *CumDist* for an MD Strength of 1100. Now check the value in the Standard Normal Table for z = 2. They are the same!

Is this an amazing coincidence? Not really. Remember that a z-value measures the number of standard deviations that an observation is from its mean. In this case, 1100 is 100 units (or 2 standard deviations) away from 1000 (the mean).

Exercise 2. What is the probability that a tissue will have an MD Strength that is equal to or less than 850? Verify that this value is the same as the z-score for z = -3.

The cumulative probability values are useful for finding probabilities to the **left** of a particular value, that is left tail probabilities, but what if we want to find the probability that a tissue will have an MD Strength that is greater than or equal to a particular value. This is not difficult if you remember that the total area under a normal curve must sum to 1. The relationship between left and right tail areas is shown in Figure 8.5.

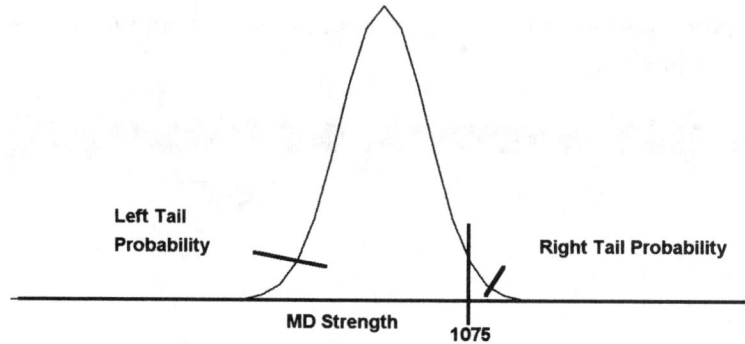

Figure 8.5 Left and Right Tails of a Normal Distribution

In order to find the probability that a tissue will have an MD Strength **greater than** a given value, simply look up the corresponding value in the *CumDist* column and subtract the number you find from 1.

Remember that if you multiply a probability by 100 you can refer to it as a percentage.

Exercise 3. What percentage of the tissues should have an MD Strength that is greater than or equal to 875?

Section 8.5   Plotting the Normal Distribution

What does the graph of this normal distribution look like? To display a graph of the distribution described in the table choose **Graph** > **Plot** from the main menu bar The **Plot** dialog box is shown in Figure 8.6.

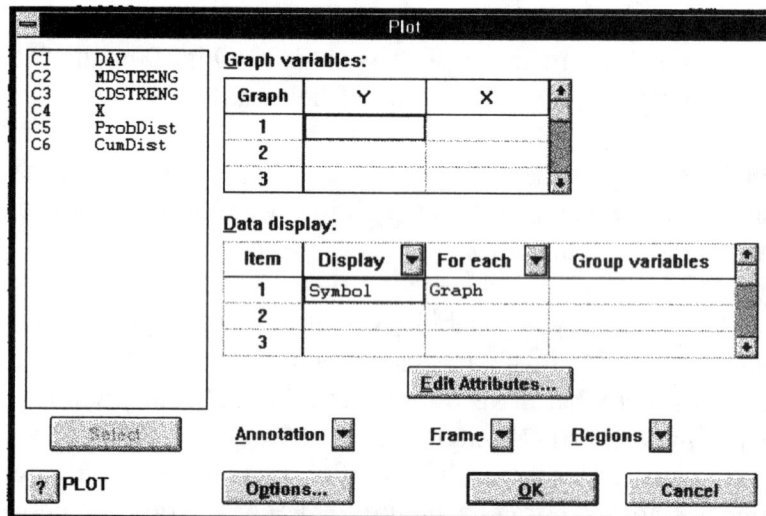

Figure 8.6 The Plot Dialog Box

Plot allows you to look at the relationship (scatter plot) between two variables. Position the cursor in the text box labeled **Y** and select the column of probability densities that you created from the list box at the left. For the **X** variable of the plot select the column containing the list of data values you created.

The standard form for a scatter plot is to simply plot a point for each pair of (x,y). Since you want a *curve*, you need to indicate that you want to *connect* the points. In the section of the dialog box for **Data display:** click on the arrow for **Display**. The pop-up menu shown in Figure 8.7 will appear.

<u>D</u>ata display:

Item	Display ▼	For each ▼	Group variables	↑
1	Symbol	Area		
2		Connect		
3		Lowess		↓
		Project		
		Symbol	tes...	

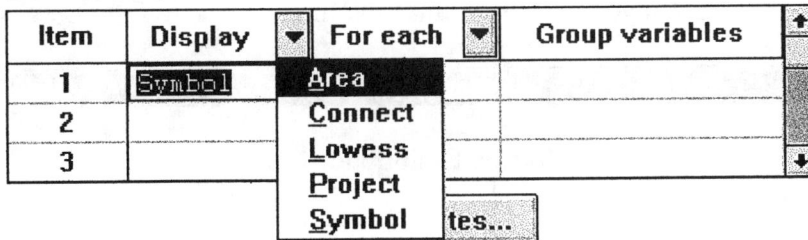

Figure 8.7 Pop-up Menu for Plot Display

Select **Connect** from the menu and click on **OK**. The plot shown in Figure 8.8 should appear.

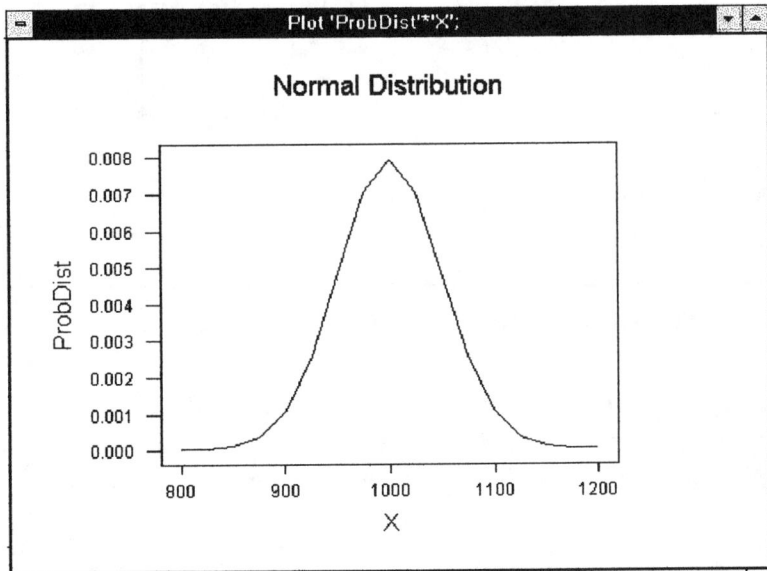

**Figure 8.8 Plot of Normal Distribution with Mean = 1000
and Standard Deviation = 50**

You may notice that the graph shown in Figure 8.8 does not really look like the normal distribution that we are used to seeing. The curve is not really very smooth. This is because you created the illusion of a curve by plotting the values of the probability density and connecting the points. Since you specified a step value of 25, there were only had 17 values to use to plot the curve. For a curve as complicated as the normal curve this is not nearly enough! You can make a much smoother curve by making the increment value smaller.

Exercise 4.   Plot a normal curve for a distribution with a mean of 1000 and a standard deviation of 50 using an increment of 10.

The graph you create should look like the one in Figure 8.9.

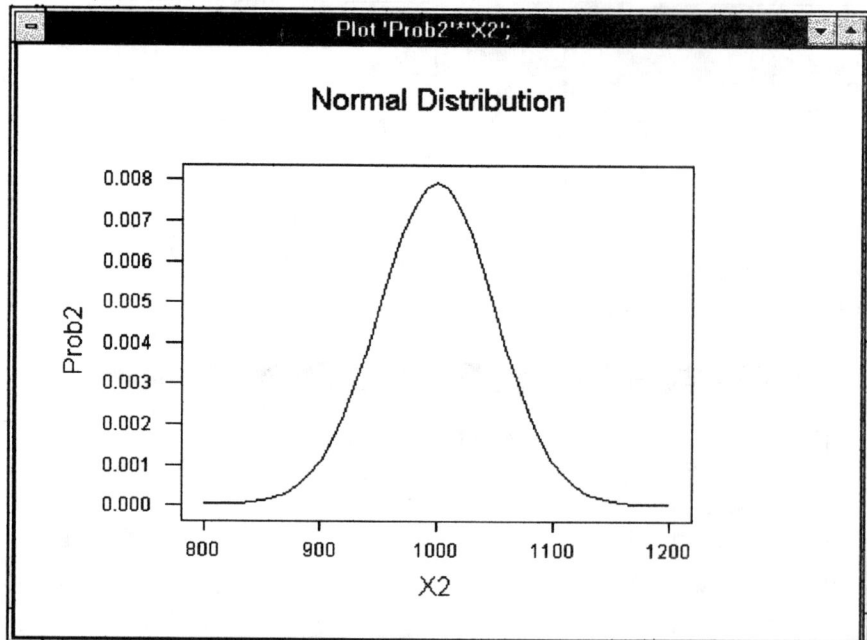

Figure 8.9 Normal Curve with Increment = 10

To figure out what values to use for START and END use equations 8.4.1 and 8.4.2

Exercise 5. Create a normal probability table and a plot of the density function for the specified values of CD Strength. (You will find the values at the beginning of this chapter)

Section 8.6   Changing the Normal Distribution

There are two parameters that determine the way a normal curve looks, the mean and the variance or standard deviation. The **mean** determines where on the number line the curve is **centered** and the **standard deviation** determines how the curve **spreads out** around the mean.

In Section 8.4 we saw that the probability that the MD Strength of the tissue was too high was .0668, while the probability that it was too low was .00135. It might be reasonable to look at the target values again and see if they can be adjusted to reduce the percentage that are out of specification. First you will look at the effects of decreasing the mean and then the standard deviation.

Exercise 6. Create probability density column for a normal distribution with a mean of 975 and a standard deviation of 50. **Use the same column of data values that you did for the MD table in Exercise 4**. (Don't make a new column of values based on the formulas!)

The first few rows of the new column are shown in C9 of the worksheet shown in Figure 8.10.

	C4	C5	C6	C7	C8	C9	
↓	X	ProbDist	CumDist	X2	Prob2	Prob3	
1	800	0.0000027	0.000032	800	0.0000027	0.0000175	
2	825	0.0000175	0.000233	810	0.0000058	0.0000345	
3	850	0.0000886	0.001350	820	0.0000122	0.0000653	
4	875	0.0003506	0.006210	830	0.0000246	0.0001191	
5	900	0.0010798	0.022750	840	0.0000477	0.0002084	
6	925	0.0025904	0.066807	850	0.0000886	0.0003506	
7	950	0.0048394	0.158655	860	0.0001583	0.0005665	

Data window.

**Figure 8.10 Probability Densities for $\mu = 975$
and $\sigma = 50$**

In order to see the real impact of the change it is best to view the two normal curves on the same graph. From the main menu bar select **Graph** > **Plot**. To plot multiple curves on the same graph you need to specify each graph separately For **Graph 1** select the probability densities for the normal curve for with $\mu = 1000$ and $\sigma = 50$ for **Y** and the corresponding set of data values for

X. Move the cursor to the line for **Graph 2** and this time select the column with the probability densities for $\mu = 975$ and $\sigma = 50$. For **X**, use the same set of data values that you used for **Graph 1**.

Unless you specify otherwise, Minitab will produce separate graphs. At the bottom of the dialog box, click on the arrow next to **Frame** and select **Multiple Graphs...** The Multiple Graph dialog box is shown in Figure 8.11.

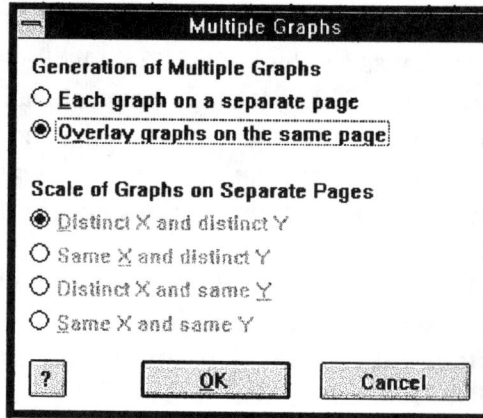

Figure 8.11 Multiple Graph Dialog Box

Click on the option button for **Overlay graphs on the same page**. Click on **OK** twice to obtain the plot shown in Figure 8.12.

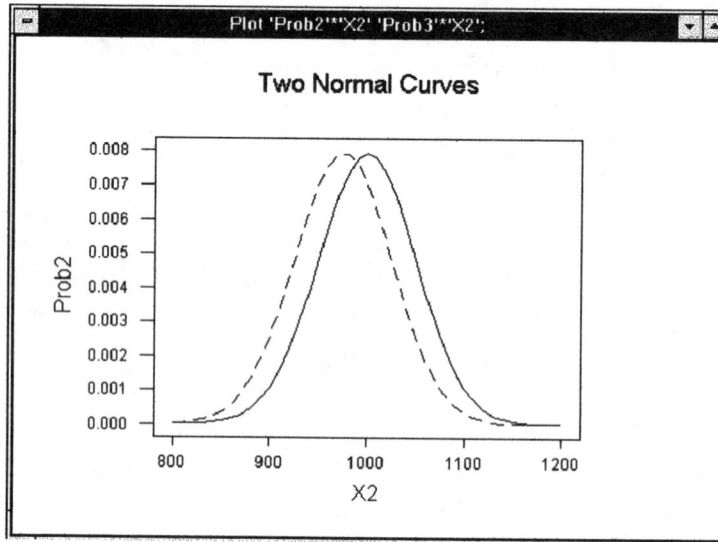

**Figure 8.12 Normal Curves with Different Means
and Same Standard Deviations.**

You can find the probabilities that the MD Strength will fall below 850 or above 1075 for the new distribution. From the main menu bar select **Calc** > **Probability Distribution** > **Normal**. To find these probabilities you want the **Cumulative probability** for each value. Since you are only interested in two values, it is easier to use **Input constant:** than **Input column:**. Click on the option button for **Input constant:** and enter 850 in the text box. Click on **OK** and the output will be displayed in the **Session** window as shown in Figure 8.13.

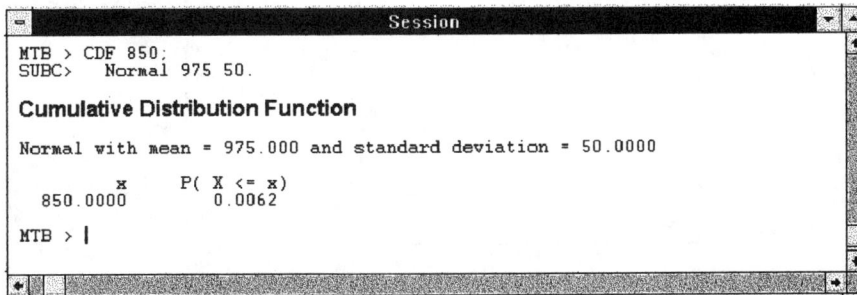

```
                              Session
MTB > CDF 850;
SUBC>    Normal 975 50.

Cumulative Distribution Function

Normal with mean = 975.000 and standard deviation = 50.0000

         x        P( X <= x)
  850.0000          0.0062

MTB > |
```

Figure 8.13 Cumulative Probability for X = 850

Did you remember that to find the right tail probabilities you must subtract the value you obtain from 1?

Repeat these steps for an X value of 1275. From the output you can see that the percentage of MD Strengths below 850 increases slightly from 0.135% to 0.621% while the percentage above 1075 decreases quite a bit from 6.68% to 2.28%. It would seem that decreasing the MD Strength specifications by 25 lb./ream will not have a negative effect on the sheets tear problem and may in fact help the softness problem.

Exercise 7. Create a column of probability densities for a normal distribution with a mean of 1000 and a standard deviation of 40. Use the of data values that you have been using. Plot the normal curve for this distribution on the same graph as the curve for the original MD Strength specifications.

The graph you just created should resemble the one shown in Figure 8.14.

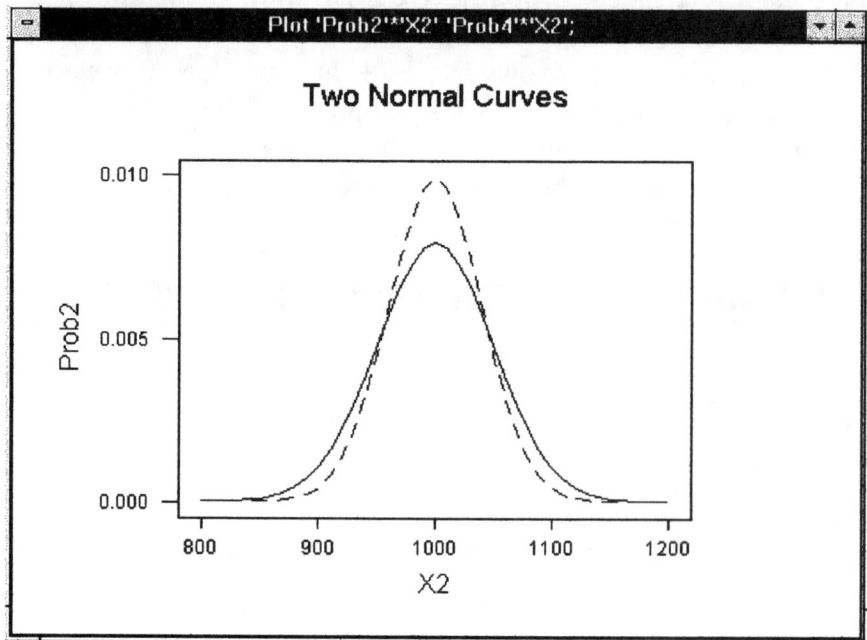

Figure 8.14 Normal Curves With the Same Mean and Different Standard Deviations.

Section 8.7   Empirical Rules for Testing Normality

The specifications assume that the MD strength measurements are normally distributed. If they are not, then even if the mean and standard deviation of the data match the specifications, the probabilities of having defective tissues will not be what you expected! Although there are more formal ways to test to see if data are normally distributed, you can use what you know about the normal distribution to make an empirical (approximate) check.

Anytime that you are examining data you should summarize the data both descriptively and graphically. These methods give you insights that you cannot get from just looking at numbers, and they enable you to understand, intuitively, the results of more sophisticated statistical tests.

Chapter 8   The Normal Distribution   175

Since you are interested in knowing whether or not the process data conforms to the specifications, you want to know if the mean and the standard deviation are correct and whether the distribution is normally distributed. In later chapters you will learn exact statistical tests to answer these questions, but right now you can get a good idea of whether or not they are correct.

The first step then, is to get a set of summary statistics for *MDStrength* and by creating a histogram of the data.

Exercise 8. Calculate a set of descriptive statistics for the data in the variable *MDStrength*.

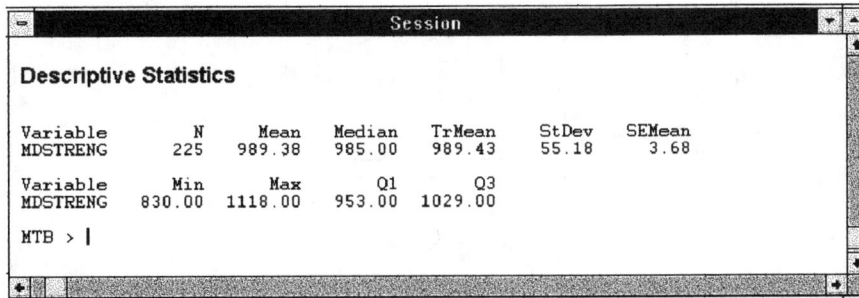

```
━                          Session                          ▲

Descriptive Statistics

Variable        N      Mean    Median   TrMean    StDev   SEMean
MDSTRENG      225    989.38    985.00   989.43    55.18     3.68

Variable      Min       Max        Q1       Q3
MDSTRENG   830.00   1118.00    953.00  1029.00

MTB > |
```

Figure 8. 15 Summary Statistics for *MDStrength*

From the descriptive statistics shown in Figure 8.15 you see that the sample mean MD Strength is approximately 989.4 and the standard deviation is approximately 55.2. While these values are not exactly the 1000 and 50 of the specifications they would appear to be close. Comparing the median which is approximately 985 to the mean it appears that the distribution of *MDStrength* is symmetric, which is one of the important characteristics of the normal distribution.

The next step is to view the data graphically to see if the assumption of normality even makes sense. To do this you will need to create a relative frequency histogram for the variable *MDStrength*.

Exercise 9. Create a relative frequency histogram for the variable *MDStrength* that uses cutpoints and goes from 800 to 1200 in class intervals of 50 units. Use data labels to display the relative frequency for each cell.

Your graph should resemble the one shown in Figure 8.16.

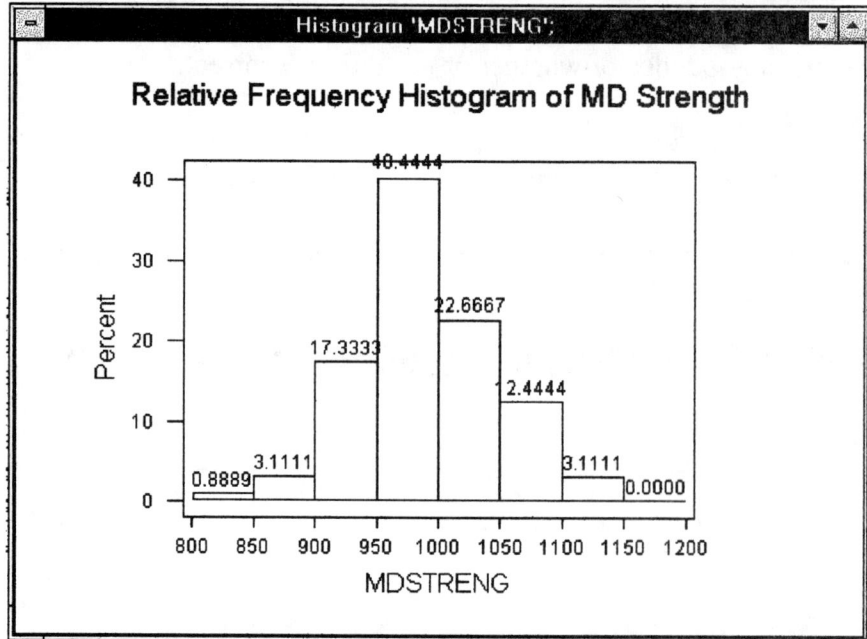

Figure 8.16 Relative Frequency Histogram for
MDStrength

Exercise 10. Examine the histogram you just created carefully. Does the data appear to be normally distributed?

The histogram shows a distribution that is symmetric and mound shaped. Thus, it would appear that the normal distribution is a reasonable assumption.

Since interpreting graphical displays of data is largely subjective it is a good idea to use another method in conjunction with the graph. The Empirical Rule can be used to determine whether the percent of data falling within 1, 2 and 3 standard deviations of the mean corresponds to what would be expected from the normal distribution. Looking at the relative frequencies on the histogram you just created, you can see that the percent of the data that are within one standard deviation (between 950 and 1050) is $40.44\% + 22.67\% = 63.11\%$. From the Empirical Rule, approximately 68% of the population should be within one standard deviation of the mean. Thus, the actual data is not too far off, but you still have to check the percents within two and three standard deviations to get the whole picture.

Exercise 11. Use the procedure above to determine the percent of *MDStrength* data that are within two and three standard deviations of the mean.
92.88

You should have gotten values of 92.88% for two standard deviations and 99.1% for three standard deviations. Comparing these to the values of Empirical Rule, 95% and more than 99% you can conclude that the assumption of normality is not too far off.

Section 8.8 Investigative Exercises

In the following exercises you are asked to use the skills introduced in the previous chapters to extract information from the tissue strength datafile. You are provided with space to answer the questions and paste in graphical output from the program. If you do not have access to a printer, you can sketch the graphs on the axes provided.

1. a) Using the relative frequency histogram that you generated in the preceding chapter, what percent of the *MDStrength* data does not meet the critical specifications? (That is, what percent are actually defective?)

b)   How does this compare to the percent defective expected by the product specifications?

c) Do you think that the company should be concerned about the difference? Why or why not?

2. According to the product specifications, CD Strength is supposed to be normally distributed with a mean of 450 and a standard deviation of 25.
a) Create a set of normal probabilities for the theoretical distribution of CD Strength.

b) The critical values for CD Strength are 480 on the high side and 390 on the low side. CD Strength that is too high creates a stiff tissue. Since the CD direction is the one in which tissues are pulled from the box, a value of CD Strength that is too low can cause sheets to tear on removal from the carton. According to the specifications, what percent of the tissues should have CD Strengths that are too high? too low?

3. Create a graph of the theoretical distribution of CD Strength.

4. Generate a set of descriptive statistics for the variable *CDStrength*.

a) Compare the mean and the median. Do you think that the distribution of *CDStrength* is symmetric?

b) Looking at the z-scores of the largest and smallest values of *CDStrength*, do you think there are any outliers?

c) By comparison, do you think that the *CDStrength* meets the values for the mean and the standard deviation that are in the product specifications?

d) Looking at the descriptive statistics, do you think that *CDStrength* is normally distributed?

5. a) Create a relative frequency histogram for the variable *CDStrength*. Does it support the assumption of normality?

b) Compare the percent within 1, 2 and 3 standard deviations to the theoretical values. Is the assumptions of normality still reasonable?

6. Another strength criteria which is mentioned in the product specifications is the **Geometric Mean Tensile (GMT).** This is not an actual measurement, but is a value calculated from MD and CD Strength. The GMT is found by taking the square root of the product of MD and CD Strength.
a) Create a new variable *GMT* in your worksheet.

b) Based on all of the criteria do you think that *GMT* is normally distributed? Include reasons for your conclusions.

7. The critical specifications for GMT are 575 on the low side and 700 on the high side.
a) What percent of the product is expected to be defective according to the theoretical specifications?

b) What percent of the actual product is defective?

8. Prepare a report to management that indicates whether or not the process appears to be running to the product specifications. In this report include any changes that need to be made (in terms of mean and standard deviation) to bring the process back to target values.

Also make any suggestions for changing the target values that you think are appropriate. You cannot change the critical (defective) values, but you might be able to adjust the target values for the mean and standard deviation that would improve the sheets tearing problem. Include all appropriate tables and graphs in the report.

Chapter 9   "Diaper Weight"

A Study of the Central Limit Theorem

Section 9.1 Overview

Statistical Objectives: After reading this chapter and doing the exercises a student will:

- Know what a sampling distribution is and why it is important.
- Know that the sample mean is normally distributed
- Know what the Central Limit Theorem means
- Know the effects of sample size on the distribution of the mean
- Know the effects of the distribution of the original data on the distribution of the sample mean.
- Know what it means for a measurement to be in statistical control.
- Know how to identify from a control chart when a measurement is in or out of statistical control.

Section 9.2 Problem Statement

Most large manufacturing companies use some form of **Statistical Quality Control** (SQC)in the manufacture of their products. One form of SQC that is often used is a **Control Chart.** A control chart looks at statistics from samples of products taken over time. The sample statistic that is often observed is the sample mean.

A large company that manufactures disposable diapers collects data from its machines at random times during the day.

One of the variables that is measured during this time is diaper weight. Diaper weight is an important factor in the manufacturing process for two reasons. The first reason is that the material that contributes most to diaper weight is the most expensive component of the diaper. Thus it is reasonable to want to provide enough of this material, but not an excessive amount. The second reason is that the weight of a diaper relates to the consumer's perception of how well that diaper will absorb liquid. Thus it is also reasonable that the amount of the material be adequate to satisfy the consumers.

When diapers are collected they are collected in samples of size 5. The sample number and the individual diaper weights and bulks are recorded. The sample averages are then calculated and plotted on charts. Machine operators use these charts to tell them whether or not the machine is behaving as expected or whether the machine needs adjustment.

Section 9.3 Characteristics of the Data Set

FILENAME: CHAP9.MTW
SIZE: COLUMN   14
                      ROWS   250

The datafile is shown in Figures 9.1 and 9.2.

	C1	C2	C3	C4	C5	C6	C7
↓	SAMPLE_	DIAPER_	WEIGHT	BULK	WGT_1	WGT_2	WGT_3
1	1	1	55.87	0.419	55.87	55.35	54.50
2	1	2	55.35	0.380	54.85	54.79	54.65
3	1	3	54.50	0.365	54.40	56.44	54.11
4	1	4	53.97	0.406	54.69	54.55	54.62
5	1	5	54.29	0.360	54.92	54.44	56.30
6	2	1	54.85	0.397	54.89	55.06	55.45
7	2	2	54.79	0.405	54.32	55.72	54.91

**Figure 9.1 First Seven Columns of the Diaper
Weight Data Set**

	C8	C9	C10	C11	C12	C13	C14
	WGT_4	WGT_5	BULK_1	BULK_2	BULK_3	BULK_4	BULK_5
1	53.97	54.29	0.419	0.380	0.365	0.406	0.360
2	55.56	55.82	0.397	0.405	0.393	0.413	0.367
3	54.67	54.56	0.397	0.345	0.421	0.419	0.399
4	54.22	54.87	0.386	0.346	0.410	0.434	0.373
5	55.00	55.68	0.425	0.404	0.400	0.399	0.396
6	55.23	55.75	0.384	0.405	0.402	0.378	0.397
7	54.40	55.78	0.413	0.386	0.398	0.376	0.392

**Figure 9.2 Last Seven Columns of the Diaper
Weight Data Set**

Notes on the Datafile: In this file, the same data appear in two
different formats. This makes it easier to use Minitab to analyze the
data.
Columns 1 to 4 record the data in one format:

1. The variable *Sample#* keeps track of the sequence in which the
samples were taken and goes from 1 to 50.

2. The variable *Diaper* keeps track of the individual diapers within
a sample and goes from 1 to 5.

3. The variable *Weight* records the diaper weight in grams.

4. The variable *Bulk* records the diaper bulk in millimeters.

Columns 5 to 14 record the same data in another format:

5. The variables *Wgt_1* through *Wgt_5* record the 5 diaper
weights from a single sample, each sample in a different row.

6. The variables *Bulk_1* through *Bulk_5* record the 5 diaper bulks
from a single sample, each sample in a different row.

Load the datafile into Minitab and save it again under a different
name.

Section 9.4 Finding Sample Means for Row Data

In order to see what the machine operators plot for each sample you will have to calculate the mean for diaper weight and bulk for each sample. You will do this by using the **Row Statistics** function in Minitab.

You would like to find the average of the five observations for each sample and store those averages as a new variable in another column. From the main menu bar select **Calc > Row Statistics.** The dialog box shown in Figure 9.3 will appear.

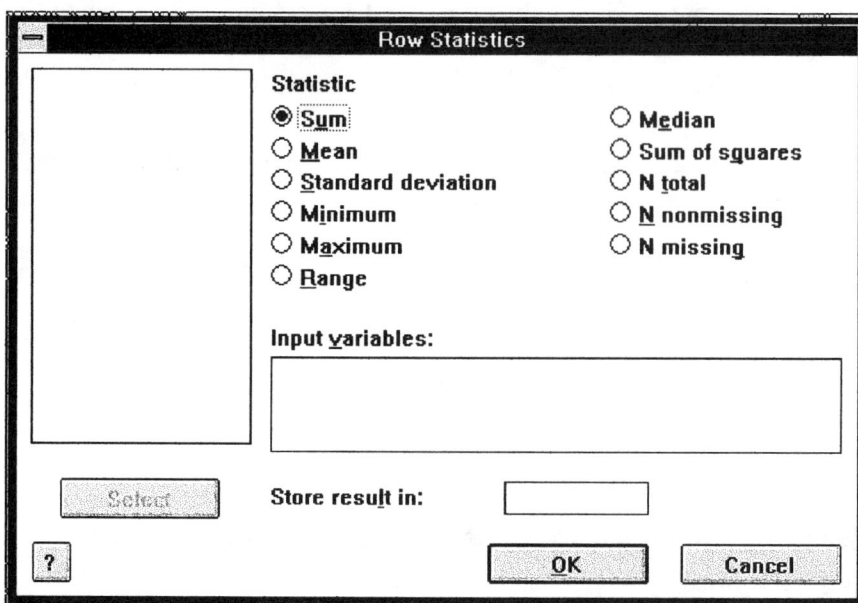

Figure 9.3 Row Statistics Dialog Box

Remember, to enter a variable or a list of variables you can type in the columns or the variable names or select them by clicking on them in the list box.

Since you want to calculate the average for each sample, click on the option button labeled **Mean**. Position the cursor in the text box labeled **Input variables:** and enter **C5-C6**. The last thing to do is to tell Minitab where you want to put the row averages. Position the cursor in the text box labeled **Store results in:** and enter **C15**, or some other empty column. The finished dialog box will look like the one in Figure 9.4.

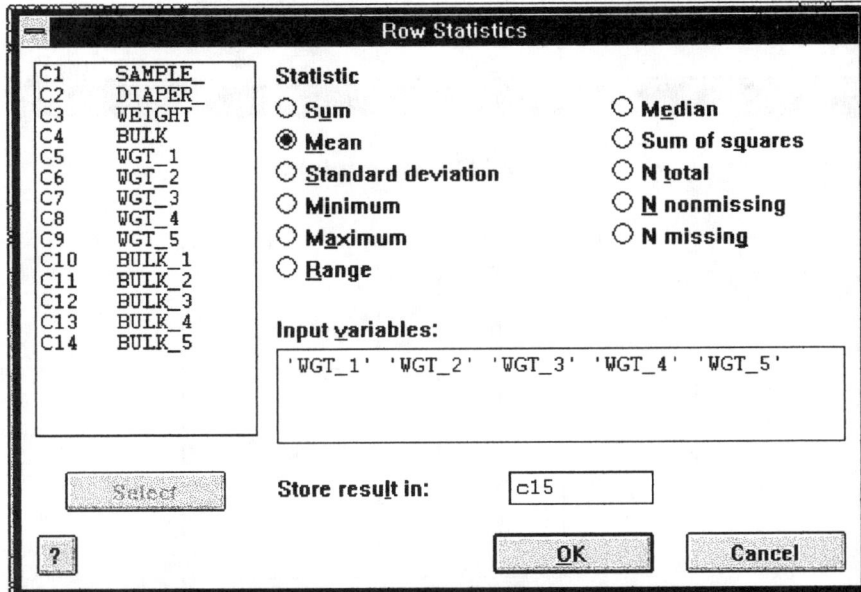

Figure 9.4 Completed Row Statistics Dialog Box

Click **OK** or press ⌨ and a new column will be added to the worksheet. Repeat the process for the variable *Bulk*. The new columns should look like the ones shown in Figure 9.5.

C14	C15	C16
BULK_5	AVGWGT	AVGBULK
0.360	54.796	0.3860
0.367	55.134	0.3950
0.399	54.836	0.3962
0.373	54.590	0.3898
0.396	55.268	0.4048
0.397	55.276	0.3932
0.392	55.026	0.3930

Figure 9.5 New Variables *AvgWgt* and *AvgBulk*

Exercise 1. Calculate a set of summary statistics, a frequency table and a histogram of the variable *Weight*.

The histogram for *Weight* is shown in Figure 9.6 and the descriptive statistics are shown in Figure 9.7.

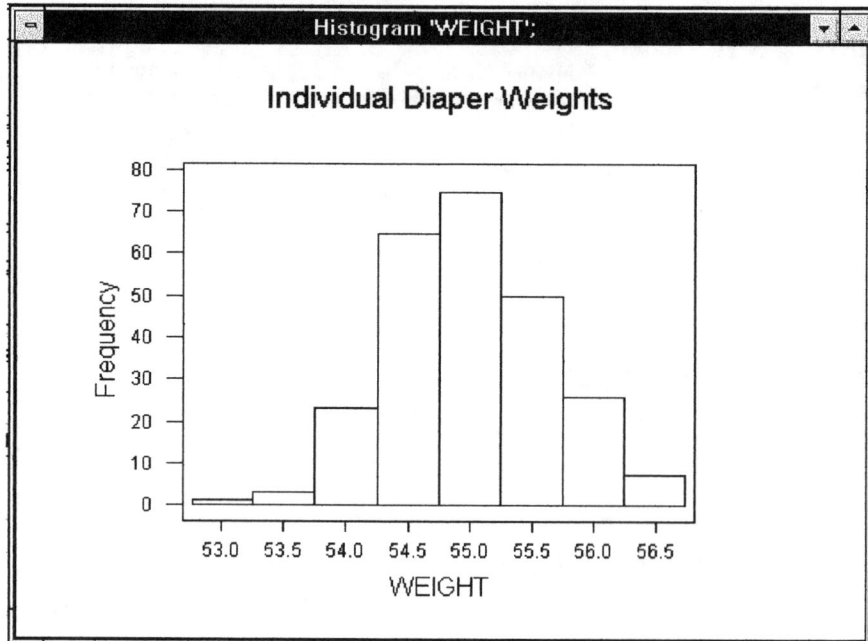

Figure 9.6 Histogram of Individual Diaper Weights

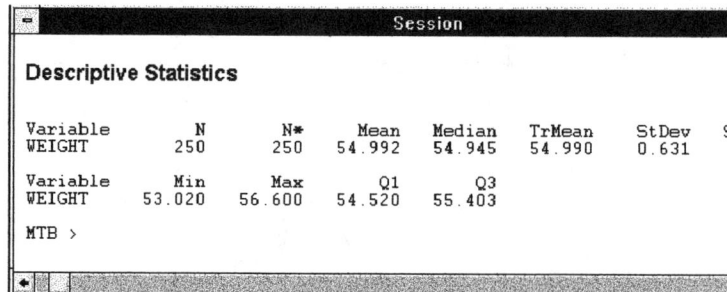

**Figure 9.7 Descriptive Statistics of Individual
Diaper Weights**

Exercise 2. Use the histogram and the descriptive statistics to
determine whether it is reasonable to believe that the variable
Weight is normally distributed.

From the shape of the histogram and a comparison of the
median and the mean it would appear that diaper weight is normally
distributed with a mean of 54.99 gm and a standard deviation of
0.625 gm.

Section 9.5 The Central Limit Theorem

The descriptive statistics for the variable *Weight* shown in Figure 9.7 indicate that the weights of the diapers vary from a minimum of 53.02 to a maximum of 56.61 gm. When the machine operators plot the data on control charts, they look at the *average* of the five diapers in a sample. How does the distribution of the averages compare to the distribution of the individual diaper weights?

Exercise 3. Create a set of summary statistics, a frequency table and a histogram for the variable *AvgWgt*.

From the histogram in Figure 9.8 we see that while *AvgWgt* centers at the same value as *Weight*, it exhibits much less variability.

Remember! In order to make a good visual comparison use the same beginning and ending values for the class intervals that were used in the histogram for the individual weights. The class widths can be different.

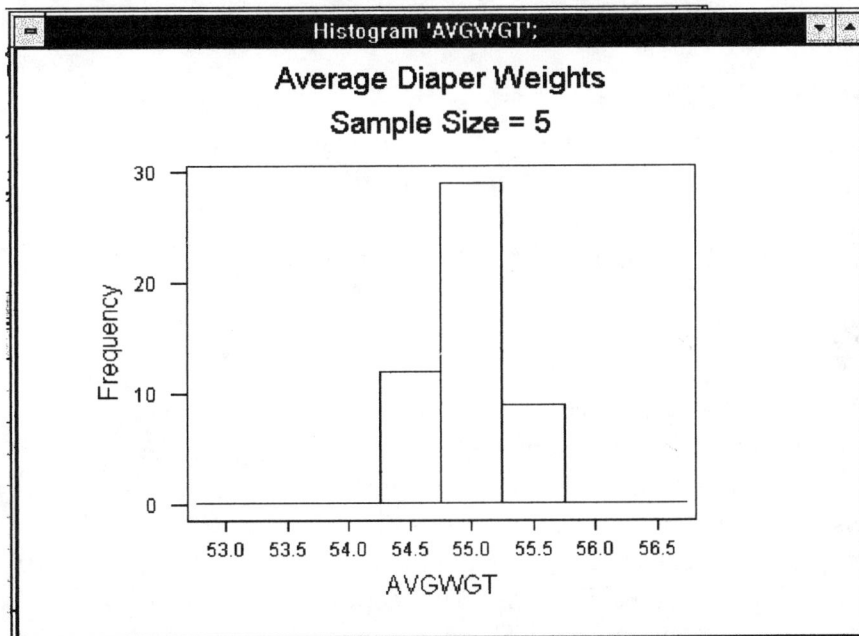

Figure 9.8 Histogram of Sample Averages for Diaper Weight

In fact, looking at the descriptive statistics shown in Figures 9.7 and 9.9 the standard deviation of the individual diapers is about 2.2 times that of the sample averages!

If you remember the Central Limit Theorem you know that for a random variable X, with mean μ and standard deviation σ, the distribution of the sample means (sampling distribution of $\overline{X}$) is normal with mean μ and standard deviation $\frac{\sigma}{\sqrt{n}}$ (standard error of the mean). This is true *exactly* when the distribution of X is normal and *approximately* when the distribution of X is not normal for large enough values of n. If you consider that the sample size used for *AvgWgt* was n = 5, and that $\sqrt{5} = 2.236...$ then this is *exactly* what you should expect according to the Central Limit Theorem.

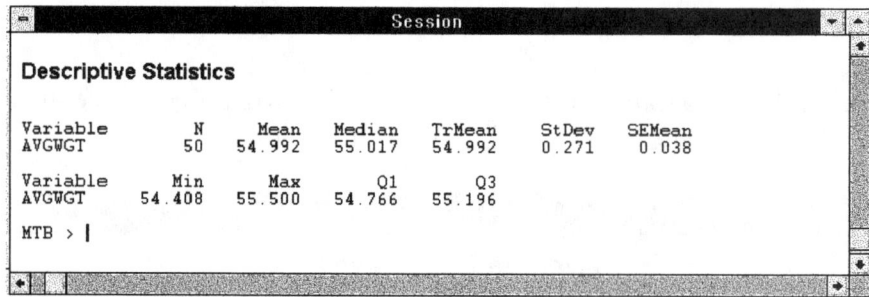

```
                                    Session

Descriptive Statistics

Variable          N       Mean    Median    TrMean     StDev    SEMean
AVGWGT           50     54.992    55.017    54.992     0.271     0.038

Variable        Min        Max        Q1        Q3
AVGWGT       54.408     55.500    54.766    55.196

MTB > |
```

**Figure 9.9 Descriptive Statistics for Average
Diaper Weights**

Section 9.6 Control Charts in Minitab

Section 9.6.1 About Control Charts

In the problem statement, you read that the machine operators plot the sample averages on *Control Charts*. A control chart is a statistical tool for determining when a random variable associated with a manufacturing process is "out of control". "Out of control" means that the measurement of interest is not behaving according to its specifications. The theory of the control chart is based on the *Central Limit Theorem*, since what is plotted is the sample average!

A control chart is a graph that plots the sample averages in chronological order along the X axis. There are three guidelines on a control chart that tell the operator whether the process is in control.

- The **Center Line** (CL) of the chart is drawn at the theoretical or specified value for the mean of the distribution.

- The **Upper Control Limit** (UCL) of the chart is a horizontal line drawn at a value that is **three standard errors of the mean** (called 3 sigma limits) above the center line.

- The **Lower Control Limit** (LCL) of the chart is a horizontal line drawn at a value that is **three standard errors of the mean** (called 3 sigma limits) below the center line.

Control limits at $\pm$ 3 sigma limits indicate events that are highly unlikely to occur if the theoretical distribution of the measurement is correct. Remember from your study of the normal distribution that 99.73% of all values from a normal distribution fall within 3 standard deviations of the mean. Thus, the probability that a sample average will fall outside the control limits is .0027 or 27 in 10,000!

As long as the measurements plotted fall within the 3 sigma limits, the process is assumed to be in control and allowed to run. If, however, a point falls outside the limits, either above the UCL or below the LCL, the process is stopped and the operators search for the cause of the erroneous value.

$$UCL = \mu + 3\frac{\sigma}{\sqrt{n}}$$

$$LCL = \mu - 3\frac{\sigma}{\sqrt{n}}$$

The control limits for a variable can be based on the specifications or based on sample data. For Diaper Weight the target mean is 55 gm. with a target standard deviation of 0.55 gm.

Using these values, if you substitute in the formulas you get:

$$UCL = 55 + 0.738 = 55.738$$
$$\text{and}$$
$$LCL = 55 - 0.738 = 54.262.$$

When a machine operator looks at the control chart they look to see if any points (*AvgWgt*) fall outside the control limits. If they do, then they conclude that the process average is not in control, that is, it is not at the value indicated by the center line of the chart. They then take action to bring the process back into control.

Section 9.6.2 Creating a Control Chart in Minitab

Minitab has options available that create control charts for a variety of different situations. To create a control chart choose **Stat > Control Charts > Xbar...**   The dialog box for the Xbar chart is shown in Figure 9.10.

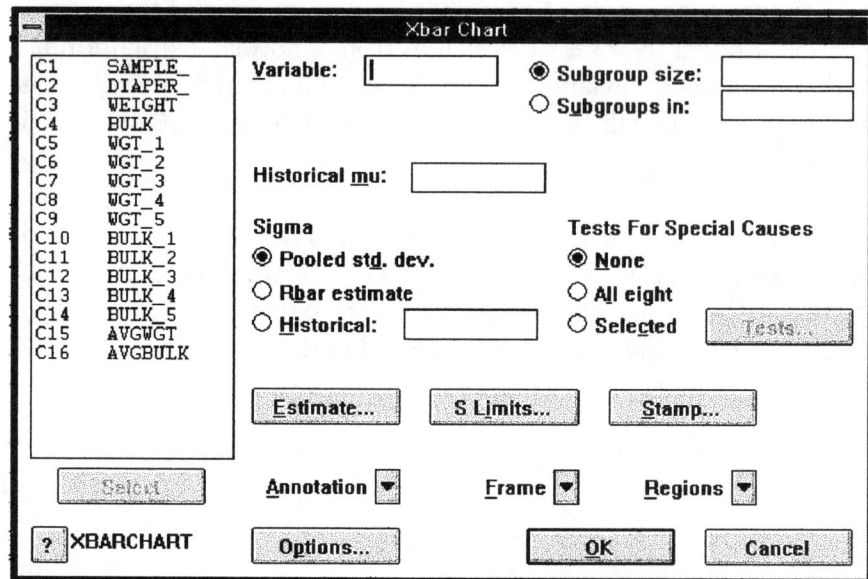

Figure 9.10 Dialog Box for Xbar Chart

Minitab expects to use the raw data to create the control chart. The data should be located in a single column. A second, optional column can be used to identify the sample number for each sample.

Position the cursor in the text box labeled **Variable:** and select *Weight* from the list box. Click on the command button for **Subgroups in:** and select *Sample* from the list box.

Next you need to indicate whether to use historical data for the control limits (e.g. specifications, data from past studies...) or to have Minitab create the limits from the data that is being charted. Position the cursor in the text box labeled **Historical mu:** and enter 55. Next, in the section of the dialog box labeled **Sigma** click on the command button for **Historical:** and enter 0.55 in the text box. Enter an appropriate title for the chart and click on **OK** to produce the chart. The control chart should look like the one in Figure 9.11.

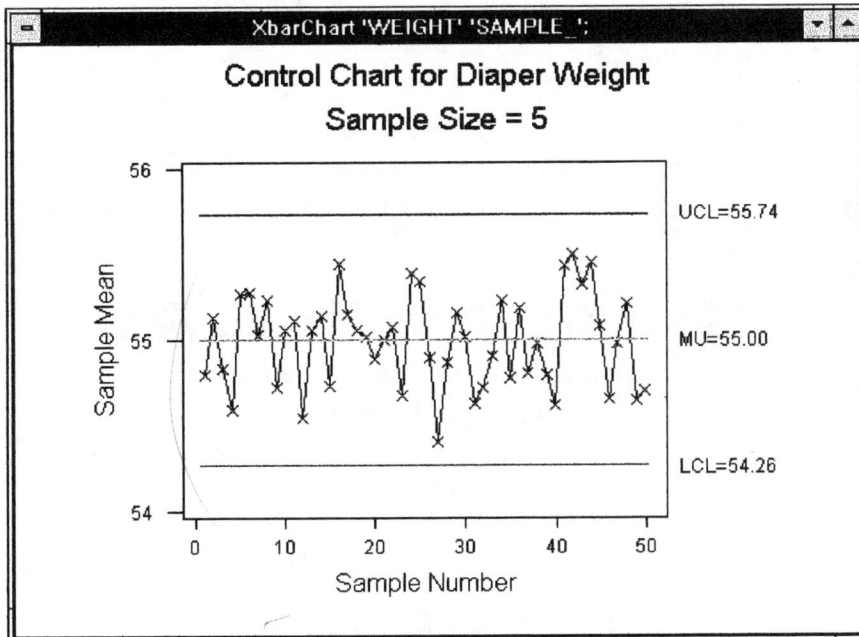

Figure 9.11 Control Chart for Diaper Weight

If you compare the values for the UCL and the LCL on the chart to the ones we calculated earlier you will see that they are the same. To use the control chart to find the standard error of the mean you have to manipulate the formulas a little. Since

$$UCL = \mu + 3\frac{\sigma}{\sqrt{n}}$$

to find $\frac{\sigma}{\sqrt{n}}$ simply subtract the value of the Center Line (MU) from the value of the UCL and divide by 3. That is,

$$\frac{\sigma}{\sqrt{n}} = \frac{55.74 - 55}{3} = 0.247.$$ This is very close to the value from the descriptive statistics for the variable *AvgWgt*. The differences are due to the method used to calculate σ from the data.

Exercise 4. Create the same control chart using the command button for **Subgroup size:** and a sample size of 5. Verify that it is the same as the one in Figure 9.11.

*Note! Use the Rbar option for calculating σ. To use the pooled standard deviation you **MUST KNOW** that the process is in control.*

Exercise 5. Create a control chart for Diaper Weight using the data to calculate μ and the value of σ. Compare this chart to the one created using the specification limits. Are they very different? What does this imply?

If you compare the chart shown in Figure 9.12, which was created using the sample data, to the one based on specifications, you see that the values for the Center Line and the UCL and LCL are not very different.

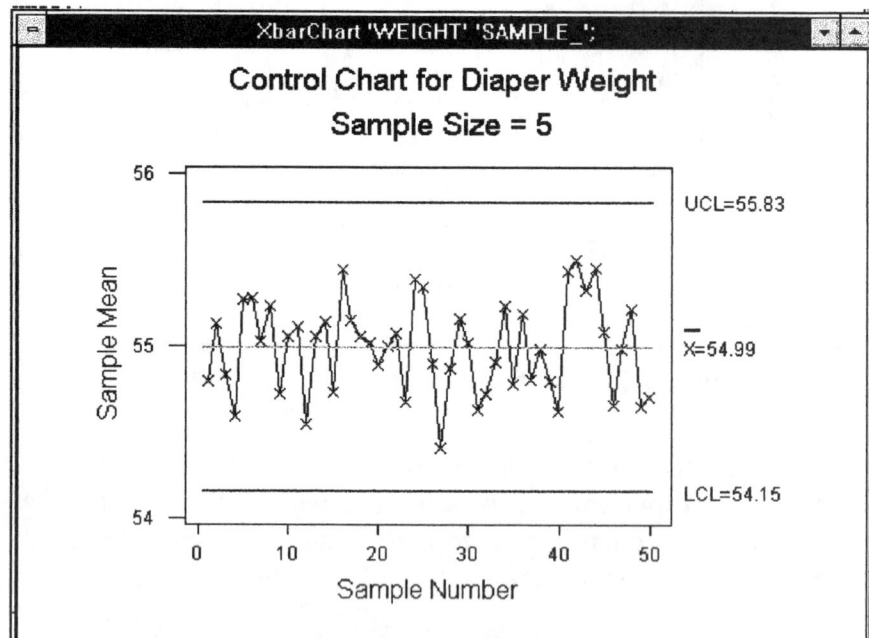

Figure 9.12 Control Chart Using Data to Calculate Limits

This seems to indicate that the actual process values do not differ very much from the specifications. In particular, the values for mu are the same and the value used for sigma from the data might be a little larger than the specification.

Section 9.7 Investigative Exercises

In the following exercises you are asked to use the skills introduced in the previous chapters to extract information from the tissue strength datafile. You are provided with space to answer the questions and paste in graphical output from the program. If you do not have access to a printer, you can sketch the graphs on the axes provided.

1. The target values for Diaper Weight are a normal distribution with a mean of 55 and a standard deviation of 0.55. Create plots of the theoretical distributions for *Weight* and *AvgWgt*. How do they compare to the histograms created from the actual data?

2. a) Examine the effects of a change in the sample size on the sampling distribution of the average diaper weight. Ignore the variable *Sample#* and create control charts for subgroups of 10, 15 and 20 diapers. Use the **data** to calculate the values for the Center Line and for the UCL and LCL. Fill out the table below:

Sample Size	Center Line (MU)	Standard Error of the Mean - UCL - MU)/3
10		
15		
20		

b) What do you notice about the values for the Center Line? Why does this happen?

c) What effect, if any, does a change in the sample size have on the distribution of *AvgWgt*?

d) How do the values for the standard errors compare to the actual values according to the Central Limit Theorem?

3. a) Calculate a set of summary statistics and a histogram for the variable *Bulk*. Does the data appear to be normally distributed?

b) Find the *AvgBulk* for the samples of size 5. (Use the data where the samples are arranged in rows.) Calculate summary statistics and make a histogram for *AvgBulk*. How do they compare to that of *Bulk*? If the target values for Bulk are a mean of 0.400 with a standard deviation of 0.02. how do they compare to the values you would expect from the Central Limit Theorem?

c) Vary the sample size for the *AvgBulk* and create control charts to fill in the table below:

Sample Size	Center Line (MU)	Standard Error of the Mean - (UCL -MU)/3
10		
15		
20		

d) How do the standard errors of the sample means compare to the distribution for the variable *Bulk*? Is this what you would expect? Why or why not?

d) What is the effect of changing the sample size for the variable *AvgBulk*? Is it similar to the effect for *AvgWgt*?

4. Look at the control chart for *AvgBulk*.
Does the measurement *AvgBulk* appear to be in statistical control? Why or why not?

5. The product specifications for the diapers include a variable *Density* which is found by dividing the *Weight* of a diaper by its *Bulk*.

a) Create the variable *Density,* calculate a set of summary statistics for it and make a histogram of the data. (Use the data from the first columns where the samples are in the columns.) Does this variable appear to be normally distributed?

b) Create control charts for *AvgDensity* using samples of size 5. How does *AvgDensity* compare to *Density*?

c) Is there a correspondence in the appearance of the control charts for Diaper Weight, Diaper Bulk and Diaper Density? Would you expect there to be one? If so why? If not, why will they disagree?

6. Using the techniques of earlier exercises, investigate the effects of changing the sample size on the distribution of *AvgDensity*.

Chapter 10 "Are We In Control?"

Hypothesis Tests - One Population

Section 10.1 Overview

Statistical Objectives: After reading this chapter and doing the exercises a student will:

- Know how to set up a and perform a hypothesis test involving a sample mean.
- Know when to use a Z-test and when to use a t-test.
- Know what the level of significance of a hypothesis test is and how to choose an appropriate level for a specific test.
- Know what it means to reject the null hypothesis.
- Know how to set up and perform a hypothesis test involving a sample variance.
- Know how to choose between the sample and theoretical variance.

Section 10.2 Problem Statement

In Chapter 8 we saw that the management of the tissue company was concerned about customer complaints involving sheets tearing on removal. They decided to look at the manufacturing process to see how it compared to the product specifications and to see if any changes needed to be made.

So far the engineers looking at the process has been able to get a good picture of what the actual process is doing and compare that, visually and empirically, to the specifications. Now they need to know whether they should adjust the process. Making changes to the manufacturing process is a BIG job and before they proceed they would like to be a little more certain that the changes need to be made. They decide that they will perform hypothesis tests on the data they have collected to see if the adjustments are really necessary.

Section 10.3 Characteristics of the Data Set

FILENAME: CHAP10.MTW
SIZE: COLUMN   3
               ROWS   200

The first seven lines of the datafile is shown in Figure 10.1.

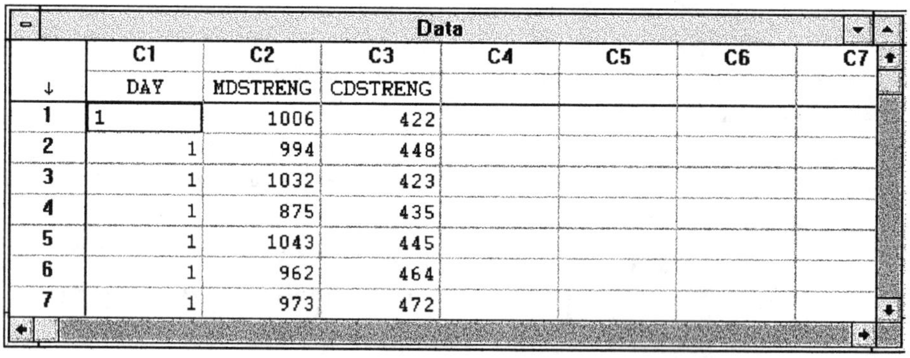

	C1	C2	C3	C4	C5	C6	C7
↓	DAY	MDSTRENG	CDSTRENG				
1	1	1006	422				
2	1	994	448				
3	1	1032	423				
4	1	875	435				
5	1	1043	445				
6	1	962	464				
7	1	973	472				

Figure 10.1 The Tissue Strength Datafile

Notes on the Datafile:

1. The variable *Day* keeps track of the day on which the sample was taken and goes from 1 to 3.

2. The variable *MDStrength* measures Machine-directional Strength and is measured in lb./ream.

3. The variable *CDStrength* measures Cross-directional Strength and is measured in lb./ream.

Load the datafile into Minitab and save it again under a different name.

Section 10.4 Testing a Hypothesis

In Chapter 8 you speculated about whether the sample data from the tissue process indicated that the process was meeting specifications. If the company is going to make a decision about whether or not to adjust the process or to change the specifications, they will need a little more than just speculation!

In order to decide whether the sample data for *MDStrength* meets the specifications we will have to **test a hypothesis.** The first step in testing a hypothesis is deciding what the appropriate hypothesis is. The product specifications for MD Strength state that the measurement is normally distributed with $\mu = 1000$ and $\sigma = 50$. You want to decided whether your sample came from a population with these parameters. That is, you want to decide whether there is any evidence that the sample data does not match the hypothesized product specifications. We can then state the hypothesis for the test as:

Null Hypothesis (H$_O$) $\mu = 1000$

Alternative Hypothesis (H$_A$) $\mu \neq 1000$

After setting up the hypothesis you need to choose a level of significance, α, for the test. For this example you will test at a 0.05 level of significance. Section 10.7.1 discusses the effects of the choice of significance level. The next step is to decide which type of hypothesis test to use, Z or t. Usually decisions on the type of hypothesis test about a mean are based on two things; sample size and whether you are using a sample or population variance. Table 10.1 lists the criteria for each of the tests.

Population	Variance	Sample Size	Test
Normal	Known	$n \geq 0$	Z - test
Normal	Unknown	$n \geq 30$	Z - test
Normal	Unknown	$n < 30$	t - test
Non-normal	Known	$n \geq 30$	Z - test
Non-normal	Unknown	$n < 30$	non-parametric

Table 10.1 Criteria for Z and t tests

For your test you have 200 observations in the sample. From

the table you can see that when $n \geq 30$ and you are using a known (hypothesized) variance the test of choice is the Z - test regardless of the distribution of the population. In this case, your work in Chapter 7 indicated that the assumption of normality did not appear to be too far offbase.

Section 10.5 Hypothesis Testing in Minitab

From the main menu bar select **Stat** > **Basic Statistics**. The Basic Statistics submenu shown in Figure 10.2 appears.

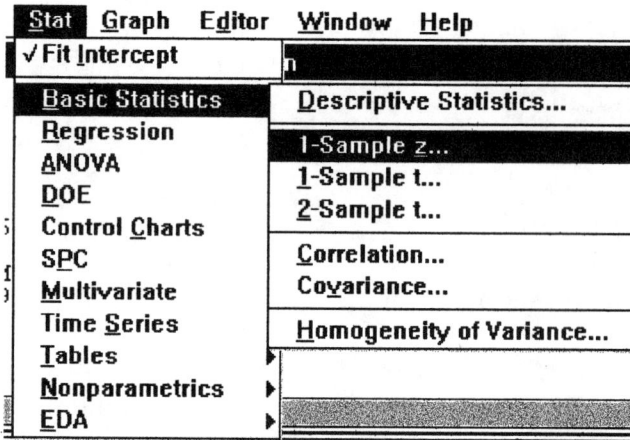

Figure 10.2 The Basic Statistics Submenu

Now you must decide what class of hypothesis tests you want to perform. The sample you want to test came from a single population, so you will be doing "One Sample" tests. Since you have already determined that you want to do a Z -test, select **1-sample z...** from the menu. You will see the dialog box shown in Figure 10.3.

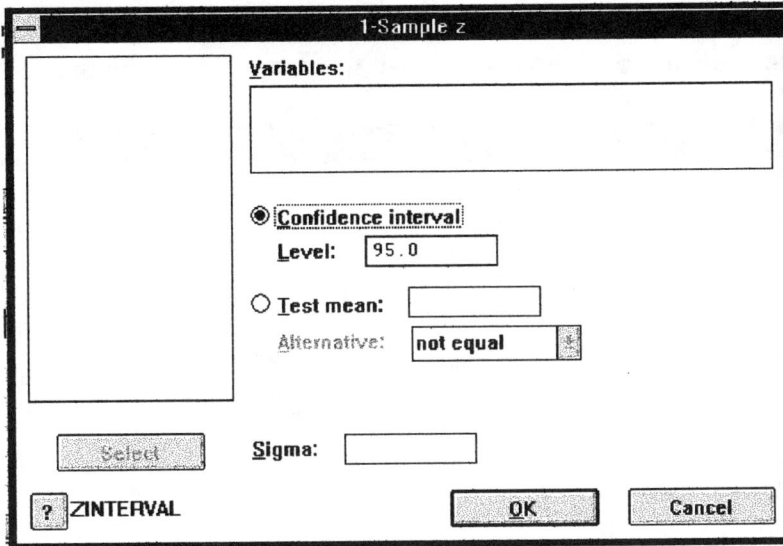

Figure 10.3 The 1-Sample z Dialog Box

Position the cursor in the **Variables:** text box and select the variable *MDStrength* from the list that appears. The **1-Sample z** option allows you to obtain a confidence interval for the population parameter of interest or to perform a hypothesis test. Click on the option button for **Test mean:**. In the text box to the right of the command button you enter the hypothesized value for the population mean μ, in this case 1000. The text box labeled **Alternative:** has an arrow for the drop down list shown in Figure 10.4. From this list you can select the form of the alternative hypothesis you want to use.

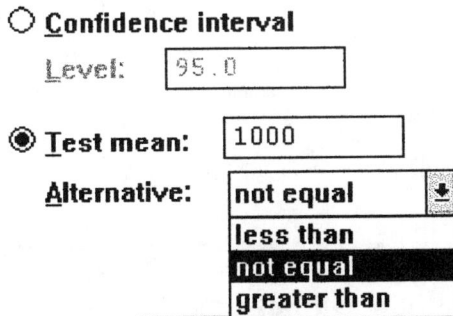

Figure 10.4 Choice for Alternative Hypothesis

The last thing you need to enter is the value for the population standard deviation, σ. Position the cursor in the text box labeled

Sigma and enter the value of σ for this test, which is 50. The completed dialog box should look like the one shown in Figure 10.5.

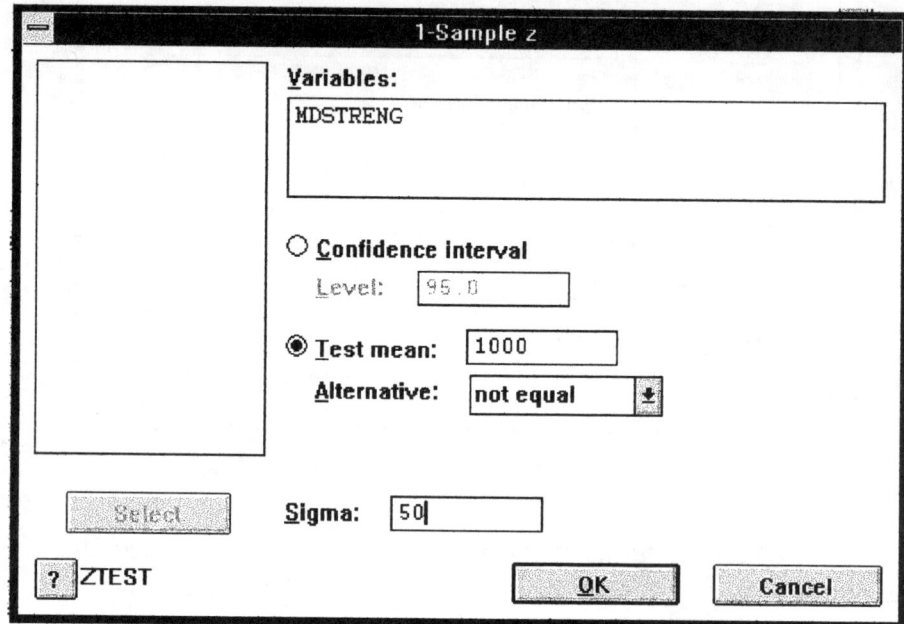

Figure 10.5 Completed 1-Sample z Dialog Box

Click on **OK** to perform the test. The results of the **1 Sample z** test will appear in the **Session** window. The results for the test you just did are found in Figure 10.6.

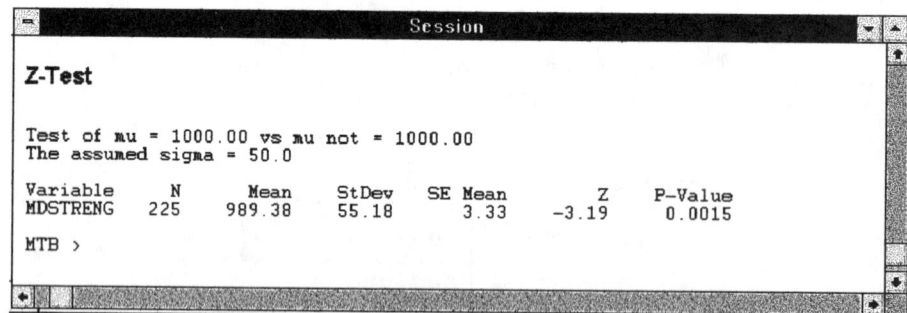

Figure 10.6 Output from 1-Sample z Test

Section 10.6 Interpreting the Test Output

Once the mechanics of the test procedure are completed you need to look at the results and make a decision about your hypotheses. The first step is to examine the output table carefully to find the values necessary to make the decision.

There are two ways to make the decision in a hypothesis test. When you do it "by hand" you look at the value of the z statistic and compare it to the appropriate critical value(s). If the z statistic is outside the critical value then you will reject H_O and conclude that H_A is true. If the z statistic is inside the critical value then you cannot reject H_O, that is, there is no evidence that it is false.

Reminder! ***Outside*** *the critical value means that the test statistic is greater than the positive critical value or less than the negative critical value*

From Figure 10.6 you see that the z statistic for the sample data is -3.19. To make the comparison you need to know the critical values for the level of significance, α, of the test. Remember that our choice was 0.05.

The critical values can be obtained from a z table (see your statistics textbook) or from Minitab. To use Minitab to find the critical value select **Calc > Probability Distributions > Normal** from the main menu bar. In the **Normal Distribution** dialog box click on the option button for **Inverse cumulative probability.** This option will give you the z value associated with a cumulative probability between 0 and 1. Since you are using the z distribution, the mean and standard deviation will be the defaults, 0.0 and 1.0.

Figure 10.7 shows that for a two sided test with $\alpha = 0.05$ the lower critical value has a tail probability (and cumulative probability) of 0.025, while the upper critical value has a tail probability of 0.025 and a cumulative probability of 1 - 0.025 = 0.975.

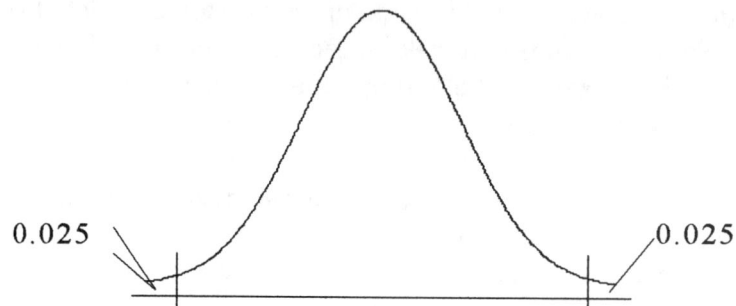

**Figure 10.7 Critical Values of the Hypothesis
Test**

Since the normal distribution is symmetric it will not matter
which critical region you enter. Click on the option button labeled
Input constant and enter either 0.025 or 0.975 in the text box. The
dialog box will look like the one in Figure 10.8.

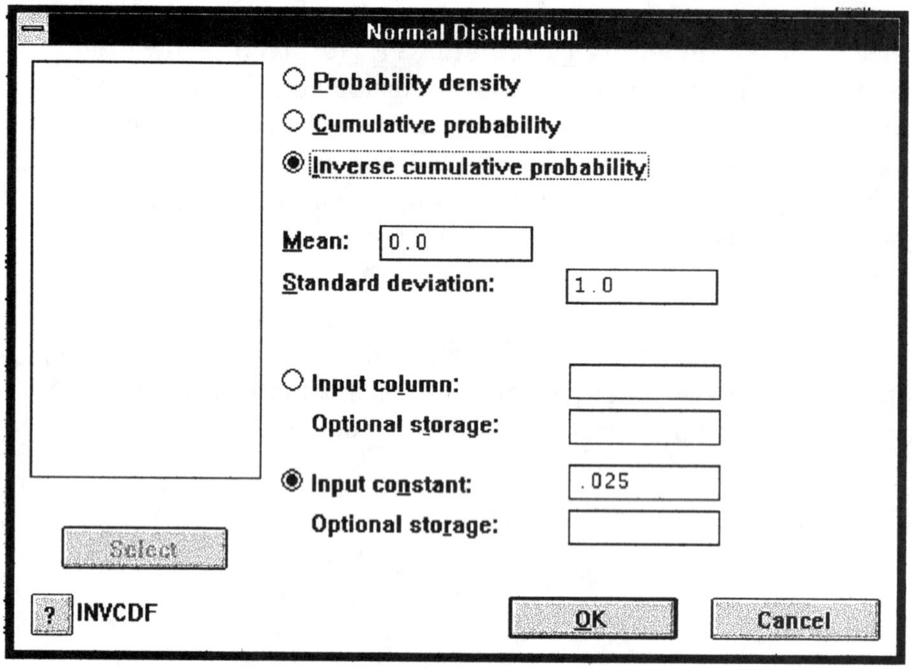

**Figure 10.8 Dialog Box for Obtaining Critical
Values**

Click on **OK**. The results will appear in the **Session** window and will look like Figure 10.9.

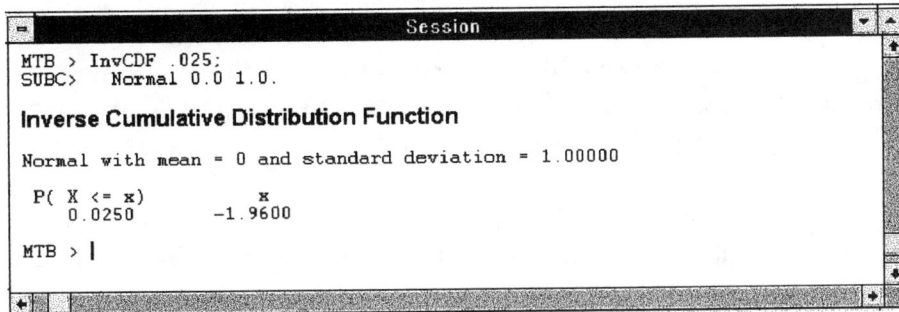

```
┌─────────────────────────────── Session ────────────────────┐
│ MTB > InvCDF .025;                                          │
│ SUBC>   Normal 0.0 1.0.                                     │
│                                                            │
│ Inverse Cumulative Distribution Function                   │
│                                                            │
│ Normal with mean = 0 and standard deviation = 1.00000      │
│                                                            │
│   P( X <= x)          x                                    │
│     0.0250        -1.9600                                   │
│                                                            │
│ MTB > |                                                    │
└────────────────────────────────────────────────────────────┘
```

**Figure 10.9 Lower Critical Value for a Two
Sided Test With $\alpha = 0.05$**

Thus for a two-sided test the critical value(s) for a two sided test, are ± 1.9600. Because -3.19 is less than the lower critical value, you reject H_O and conclude that the sample data did not come from a normal population with a mean of 1000 and a variance of 2500.

Along with the test statistic and the critical values for the test you will see that the output table in Figure 10.6 contains something called the "p-value" of the test. The p-value of the test is another way to make the decision about H_O. You may remember that the p-value is also known as the observed significance level of the test. That is, it is the smallest value of α for which the test would have resulted in rejecting H_O with the given data set. The p-value for this test is .0015. This value is the point at which the decision about H_O changes from reject to accept. Compare the value of alpha that you used for the test to the p-value. If the p-value is less than the value of alpha you will reject H_O. If the p-value is greater, you will fail to reject H_O.

Section 10.7 After the Test - Now What?

Section 10.7.1 Changing the Alpha Level of the Test

When you set up a hypothesis test you select a value for α, the significance level of the test. Recall that α, also known as the Type I error, is the probability that you will reject H_O when it is in fact true. Most textbooks define α but spend little time talking about how you go about choosing an appropriate value.

In choosing a value for α you must consider the **cost** of making a wrong decision. That is, what will you do if you reject H_O and what will it cost you (in dollars, manpower, goodwill etc.) if you are wrong.

The most common choice for α is 0.05. This is the level that most people want to use when doing hypothesis tests, but really every time you set up a test you should think about the ramifications of a wrong decision in that particular situation.

For the manufacturing company in question the results of incorrectly rejecting H_O (saying the average MD Strength is not 1000 when in fact it is) might take one of two forms. One result is that the machine operators will make adjustments to raise the level of MD Strength, since on the basis of the data it appears to be low. The other is that they will attempt to troubleshoot the process to look for the causes of low MD Strength only to find they are on a wild goose chase! In either case, the effect will be machine down time and perhaps defective product.

Suppose that the company in question decides that they cannot afford machine down time. They decide that they can only afford a 1% chance of rejecting H_O in error. Changing alpha will change the critical value of the test and perhaps your decision.

Exercise 1. Find the critical values for a two sided test when $\alpha = 0.01$. Does your decision about H_O change? If so, what is the new decision?

You could answer the question directly from the p-value of the test. Compare the new level of significance, 0.01 to the p-value of 0.015. Since the p-value of 0.0015 is less than 0.01, you will reject H_O.

Exercise 2. Suppose the company decides that the cost of machine down time is not great and that they can afford a larger Type I error. Select an appropriate value for α and find the critical values. What effect does this have on the decision from the original ($\alpha = 0.05$) test?

You might wonder why someone would ever want to make the possibility of an error, in this case a Type I error, **larger.** The reason is that along with an α error which you set for each test, there is another error that is associated with the test, the Type II error, β. The β error of the test is the probability that the test will fail to reject H_0 when it is in fact false. That is, it is the probability that the experimenter will decide that there is not enough evidence to take action and maintain status quo. In the case of the tissue manufacturer, that would mean **NOT** making changes to a process that is definitely in error.

Although actually calculating the β error is a bit complicated, the fact is that its value is in direct tradeoff with α for a sample of a given size. That is, as the probability of making a Type I error decreases, the associated Type II error probability increases. Thus the person performing the test might decide to increase α so that the β error decreases.

Section 10.7.2 One Sided vs. Two Sided Tests

For the manufacturing company, the question of interest was whether or not the process was running to target specifications. That is, whether the mean MD Strength was equal to 1000 or whether it was not equal to 1000. Thus, the test you did was a two sided test.

Suppose, however, that the company did not care if the MD Strength was high, but was concerned only about low MD Strength. That is, they would only take action if the data showed that the values were too low to have come from the target population. In this case you would want to test the hypotheses

H_0: $\mu \geq 1000$

H_A: $\mu < 1000$

In order to do this with Minitab you simply select "**less than**" from the drop down list for **Alternative**.

Exercise 3. Perform the one sided test described above to determine whether the mean MD Strength is less than 1000. What is the result?

Section 10.8 Other Considerations

At this point it is important to figure out what all this might MEAN to the manufacturing company in question. Rejecting the hypothesis means that the sample was taken when the process was not running according to the manufacturing specifications. If the sample was representative of the process at any time, then the logical conclusion is that the product they are making is not what they intend.

The natural conclusion of the test is that the mean of the process is not 1000, but is, in fact some other value. Thus the company should consider adjusting the process and retest before making any decisions about changing the specifications. **HOWEVER**, before you make a suggestion like that to the management, consider taking a second look at the test you performed.

Remember that you performed a hypothesis test using population (that is, assumed) values for both the mean AND the variance. Is it possible that the population variance is not the value that you assumed. An undetected change in the population variance can make it appear that the population mean is no longer correct.

Calculate a set of descriptive statistics for the variable *MDStrength*. The information is shown in Figure 10.10.

```
┌─────────────────────────────────────────────────────────────────┐
│                            Session                                │
├─────────────────────────────────────────────────────────────────┤
│ Descriptive Statistics                                            │
│                                                                   │
│ Variable        N      Mean    Median    TrMean    StDev   SEMean │
│ MDSTRENG      225    989.38    985.00    989.43    55.18     3.68  │
│                                                                   │
│ Variable      Min       Max        Q1        Q3                   │
│ MDSTRENG   830.00   1118.00    953.00   1029.00                   │
│ MTB >                                                             │
└─────────────────────────────────────────────────────────────────┘
```

Figure 10.10 Descriptive Statistics for
MDStrength

Look at the value of the sample standard deviation and compare it to the assumed value of 50. It would appear that the sample value of 55.18 might be higher than the assumed value. What would happen if you did the test using the sample variance instead?

Exercise 4. Repeat the original hypothesis test, but change the value of sigma to match that of the sample standard deviation. Does your conclusion change?

Because you have such a large sample size the z-test was still the appropriate test. Although this particular change did not change the conclusion of the test, you can see that it is important that you be sure that you are setting up your tests using the correct values.

Section 10.9 Investigative Exercises

In the following exercises you are asked to use the skills introduced in the previous chapters to extract information from the tissue strength datafile. You are provided with space to answer the questions and paste in graphical output from the program. If you do not have access to a printer, you can sketch the graphs in the spaces provided.

1. After giving the matter some thought the engineers looking at the tissue manufacturing process wondered if the fact that the measurements were taken on three different days might matter. Since operating conditions are not always the same on any given day, they wondered if the process was off target on all three days or if it was only off on one or two days. They decided to check each day individually.

Perform a hypothesis test at the .05 level of significance to determine whether the mean MD Strength was 1000 on each of the three days separately. Use the target standard deviation of 50 and write your results in the table below:

DAY	Z Statistic	Decision (Reject H_0 / Fail to Reject H_0
1		
2		
3		

2. For each of the three days, find the sample standard deviation and redo the test using this value. Does the decision change? Fill in the table below.

DAY	Z Statistic	Does the Decision Change?
1		
2		
3		

3. Using the results of all of hypothesis tests that you have done, do you think that the company can assume that MD Strength is running to target? Why or why not?

4. Using a .05 level of significance, does the sample data indicate that the variable *CDStrength* was running according to target specifications for the three day period as a whole?
Note: CD Strength is supposed to be normally distributed with a mean of 450 and a standard deviation of 25.

b) Test to see if the use of the sample standard deviation changes the decisions.

5. Since the consumer problem that prompted the study was sheets tearing on removal, the company decides that it is really only interested in knowing whether the average CD Strength is less than the target of 450. If it is not, they will not make any adjustments. Redo the test of the previous exercise as a one-sided test to reflect this. What is the result?

6. Judging the results of the tests you have done with CD Strength, do you see any need for the group to test each of the three days separately? Why or why not? What would you expect to happen if you did test the days separately?

7. The group is divided on the subject of GMT. The operations specialists think that since the company has critical specifications for the measurement they need to test to see if it is running according to those specifications. The machine operators say that since the GMT is calculated from the other two variables which have been tested, they do not need to perform any hypothesis tests concerning this data. Which group do you agree with and why?

8.   Using the results of your analysis prepare a report to management telling them about the current tissue manufacturing process. Make recommendations to them on whether the process needs to be adjusted or whether it is running to target. Remember, if it is running to target they will be considering making changes to the specifications to reduce customer complaints about dispensing. Address this point in your report. Compare your conclusions now to the more subjective ones you put forth in Chapter 8.

Chapter 11 "Will the Wrapper Fit?"

Hypothesis Tests - Two Populations

Section 11.1 Overview

Statistical Objectives: After reading this chapter and doing the exercises a student will:

- Know how to set up the hypotheses for a two population test concerning sample means.
- Know how to decide which two sample test is appropriate for a given situation.
- Know how to do an F-test to compare two sample variances.
- Know when the assumption of equal variances is appropriate.

Section 11.2 Problem Statement

A manufacturing company that makes paper products is having trouble with their paper towel line. When a roll of paper towels is manufactured, the last step before the roll is put in a case is to wrap it with the poly wrapper that the consumer sees it in on the store shelf. A large amount of product is being scrapped because it is not being wrapped properly.

The problem appears to be that the diameters of the towel rolls are too large and the wrapper does not go all the way around and seal properly. There are many factors in the manufacturing process that would appear to affect roll diameter, however the engineers involved were not sure exactly how different machine settings really impacted roll diameter. In fact, they were not convinced that all of the factors made a difference!

The towel machine team designed a study to determine the extent that different machine settings actually had on roll diameter. Several people on the team expressed a concern that some settings which would result in a good roll diameter might have an adverse impact on another towel roll characteristics, roll firmness.

If a towel roll is not firm that will also affect wrapping, and if it is too firm consumers will perceive it as stiff and react negatively. Fixing the roll diameter problem at the expense of firmness was not an option.

The team decided to look at three machine factors:

1. **Embosser roll gap**: the mechanism that puts the pattern in the towel
2. **Draw roll gap**: the opening through which the towel material is pulled onto the winders
3. **Speed**: the speed at which the machine winds the rolls of towels.

The machine was run at different settings for each factor and rolls of towels were taken from the end of the production line. These towel rolls were measured on two characteristics, roll diameter and roll firmness.

Section 11.3 Characteristics of the Data Set

FILENAME: **CHAP11.MTW**
SIZE: **COLUMN   5**
                        ROWS   76

The first seven lines of the datafile are shown in Figure 11.0

	C1	C2	C3	C4	C5	C6	C7
↓	DRAWROLL	SPEED	EMBOSSER	DIAMETER	FIRMNESS		
1	0	0	0	5.43307	0.271667		
2	0	0	0	5.39370	0.336667		
3	0	0	0	5.47244	0.260333		
4	0	0	0	5.39370	0.261333		
5	0	0	0	5.43307	0.297000		
6	0	1	0	5.39370	0.312000		
7	0	1	0	5.47244	0.294667		

Figure 11.1 The Towel Diameter Datafile

Notes on the Datafile:

1. The variable *Drawroll* is a 0-1 variable that indicates the size of the drawroll gap. A 0 indicates that the smaller gap measurement was used, while a 1 indicates the larger gap measurement.

2. The variable *Speed* is a 0-1 variable that indicates the machine speed. A 0 indicates the slower machine speed was used and a 1 indicates that the faster machine speed.

3. The variable *Embosser* is a 0-1 variable that indicates the size of the embosser roll gap. A 0 indicates that the smaller gap measurement was used, while a 1 indicates that the larger gap measurement was used.

4. The variable *Diameter* measures the diameter of the roll of paper towels in inches.

5. The variable *Firmness* measures the firmness of the roll of paper towels on a specialized scale. Lower numbers indicate softer rolls while larger numbers indicate firmer rolls.

Load the datafile into Minitab and save it again under a different name.

Section 11.4 Hypothesis Testing - Two Populations

Section 11.4.1 Setting Up the Hypotheses
The questions that the towel team want answered are about whether certain changes to the machine settings result in changes to two towel characteristics, mean roll diameter and mean firmness. They looked at the machines under two conditions for each machine setting and sampled product from each set of conditions. Thus they are interested in testing whether the mean towel characteristic is the same for each **population** (machine setting).   You can state the hypothesis for this test as:

H_O: $\mu_1 = \mu_2$ OR $\mu_1 - \mu_2 = 0$

H_A: $\mu_1 \neq \mu_2$ $\mu_1 - \mu_2 \neq 0$

Use **Manip > Copy** *to copy the data to two separate columns*

The first item that they decided to look at was whether there was a difference in roll diameter when the machine was run at two different speeds.

Exercise 1. Create a separate column for the variable *Diameter* for each value of the variable *Speed*.

Section 11.4.2 Choosing the Correct Test

Before you rush headlong into any statistical tests, it would be wise to **look** at the data you are about to analyze so that you have a sense of what the test results are all about. Usually you should obtain a set of descriptive statistics and a graphical display of the data. This is also useful when you need to consider the assumptions in certain statistical tests. Looking at the variable *Speed* you see that it has two values 0 and 1. Each of these values identifies the population from which the associated sample was taken.

Note: The values of Diameter are a bit strange. You will have to experiment with the step sizes to get a good picture.

Exercise 2. Create a set of summary statistics for the variable *Diameter* for each of the machine speeds separately.

Exercise 3. Create a histogram of *Diameter* for each machine speed.

The summary statistics for each sample are shown in Figure 11.2.

```
┌─────────────────────────────────────────────  Session  ────────────────────────────────────┐
│ Descriptive Statistics                                                                       │
│                                                                                              │
│  Variable   SPEED        N      Mean   Median   TrMean    StDev   SEMean                      │
│  DIAMETER       0        25    5.3953   5.3937   5.3954   0.0417   0.0083                      │
│                 1        50    5.4244   5.4331   5.4304   0.0618   0.0087                      │
│                                                                                              │
│  Variable   SPEED      Min       Max       Q1       Q3                                        │
│  DIAMETER       0    5.3150    5.4724   5.3543   5.4331                                        │
│                 1    5.2362    5.5118   5.3937   5.4724                                        │
└──────────────────────────────────────────────────────────────────────────────────────────┘
```

Figure 11.2 Summary Statistics for Diameter by Machine Speed

From the statistics you can see that the average roll diameter is larger for the higher machine speed, but is it a *significant* difference (that is, one that did not happen by chance)? To determine this you will have to perform a hypothesis test.

The first step in doing the test is to determine which test, Z or t, you should do. Table 11.1 lists the criteria for each of the tests:

Populations	Variances	Sample Sizes	Test
Independent Normal	Known	n_1 and $n_2 \geq 0$	Z-test
Independent Non-Normal	Unknown	n_1 and $n_2 \geq 30$	Z- test
Independent Normal	Unknown but assumed equal	$n_1 \leq 30$, $n_2 \leq 30$	t-test (Pooled variance)
Independent Normal	Unknown and assumed not equal	$n_1 \leq 30$, $n_2 \leq 30$	Modified t-test (Behrens - Fisher problem)

Table 11.1 Table of Test Criteria

Looking at the summary statistics you can see that the sample sizes are not both ≥ 30, and so you will need to use a t-test. Looking at Table 11.1 you also see that using a t-test requires that the populations from which the samples were taken be *independent and normally distributed* and that choice of a test depends on whether the sample variances can be assumed to be equal.

To determine whether or not the data are normally distributed you can look at the histograms of the data that you just created. While this is only an approximate test, unless there are large departures from normality (very skewed or non mound shaped distributions) the test will be valid. You could also apply the empirical rule for normality (see Chapter 8 if you do not remember how to do this).

The histograms for the two speeds are shown in Figures 11.3 and 11.4. The distributions of the data are mound shaped and while they are not exactly symmetric it would appear that the assumption of normality is not violated.

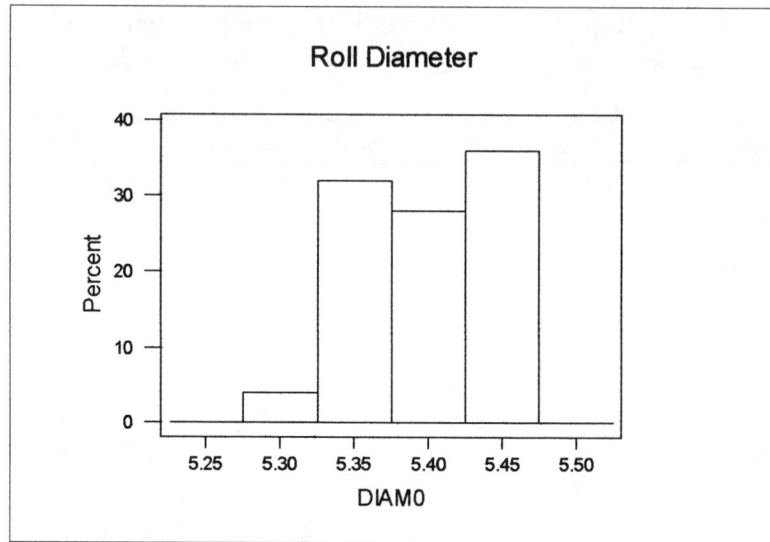

**Figure 11.3 Histogram of Roll Diameter for
Machine Speed = 0**

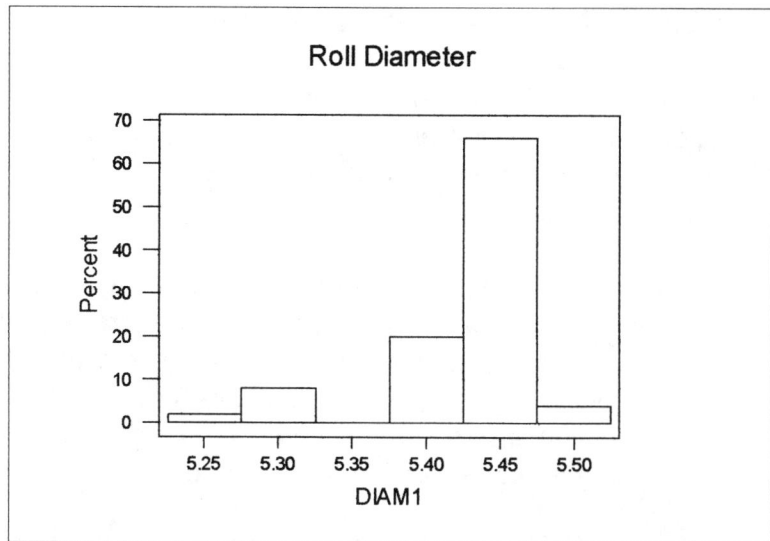

**Figure 11.4 Histogram of Roll Diameter for
Machine Speed = 1**

If you look at the data in Figures 11.2 you see that the sample standard deviation for *Diameter* with *Speed* = 0 is 0.0417 in. while the sample standard deviation for *Speed* = 1 is 0.0618 in..(The difference in variability is also evident in the histograms.) They look different but to really need to know whether they are statistically different you would have to do another statistical test. We will address this issue in the next section.

Section 11.4.3 Doing the Hypothesis Test

From the main menu choose **Stat** > **Basic Statistics** > **2-Sample t...** The dialog box shown in Figure 11.5 will open.

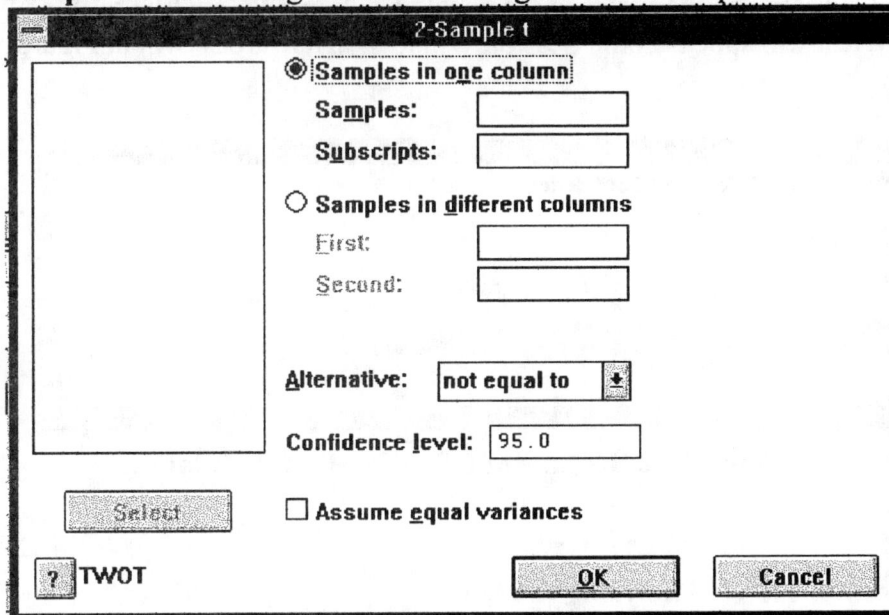

Figure 11.5 2-Sample t Dialog Box

Minitab tests the null hypothesis that the mean of the first population is equal to the mean of the second population. The data for the test can either have each sample in a separate column, or, like ours, have the variable (*Diameter*) in one column and use a second column (Speed) to define the samples. Click on the option button for **Samples in one column.** Position the cursor in the text box labeled **Samples:** and choose *Diameter* from the list. Then, position the cursor in the text box labeled **Subscripts:** and choose *Speed* from the list.

For the **Alternative:** select **Not equal to** from the drop down list. We will perform this test at the $\alpha = 0.05$ level of significance, so choose a level of confidence of 0.95.

Remember, the level of confidence is 1- α.

Now you are faced with an important decision. Remember that our choice of test depended on whether we assumed that the sample variances (or standard deviations) are equal to each other. There is a test, called an F test, to actually determine this, but it is not covered in most basic statistics courses. As a rule, unless you *KNOW* your variances are equal, it is safer to assume that they are not. The test is sensitive to this assumption and the test results that you get could be in error if they are not equal and you assume that they are. Since our standard deviations are not the same (0.0417 in. vs. 0.0618 in.) we will not assume the variance are equal. Click on **OK** and the results of the test will appear in the Session Window. They are shown in Figure 11.6.

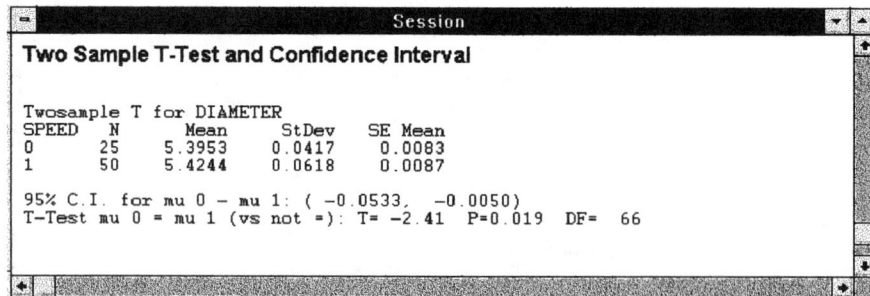

```
┌─────────────────────────────────────────────────────────────┐
│                           Session                            │
├─────────────────────────────────────────────────────────────┤
│ Two Sample T-Test and Confidence Interval                    │
│                                                              │
│ Twosample T for DIAMETER                                     │
│ SPEED   N       Mean    StDev    SE Mean                     │
│ 0      25     5.3953   0.0417    0.0083                      │
│ 1      50     5.4244   0.0618    0.0087                      │
│                                                              │
│ 95% C.I. for mu 0 - mu 1: ( -0.0533,  -0.0050)              │
│ T-Test mu 0 = mu 1 (vs not =): T= -2.41  P=0.019  DF=  66   │
│                                                              │
└─────────────────────────────────────────────────────────────┘
```

Figure 11.6 Results of t Test for Two Population Means

NOTE: Actually Minitab does not do a 2-Sample Z test. If the Z-test is the appropriate test use the t-test and DO NOT ASSUME EQUAL VARIANCES.

Section 11.4.4 Analyzing the Output

The output from the 2-Sample t test gives a little more information than the 1-sample test from the previous chapter. In addition to some summary information about each sample it also provides a 95% confidence interval for the difference between the two population means.

The results of the hypothesis test are the last line of the output. The value of the t-statistic is -2.41. In order to know what this means you must either find the critical values for the test or use the p-value of the test.

To find the critical values for a t-test choose **Calc** > **Probability Distributions** > **T...** and fill in the dialog box. Click on the option button for **Inverse cumulative distribution.** In the text box labeled degrees of freedom enter 66 (you will find this number in the last line of the test output). Last, click on the option button for **Input** constant and enter 0.025 or 0.975. The completed dialog box is shown in Figure 11.7.

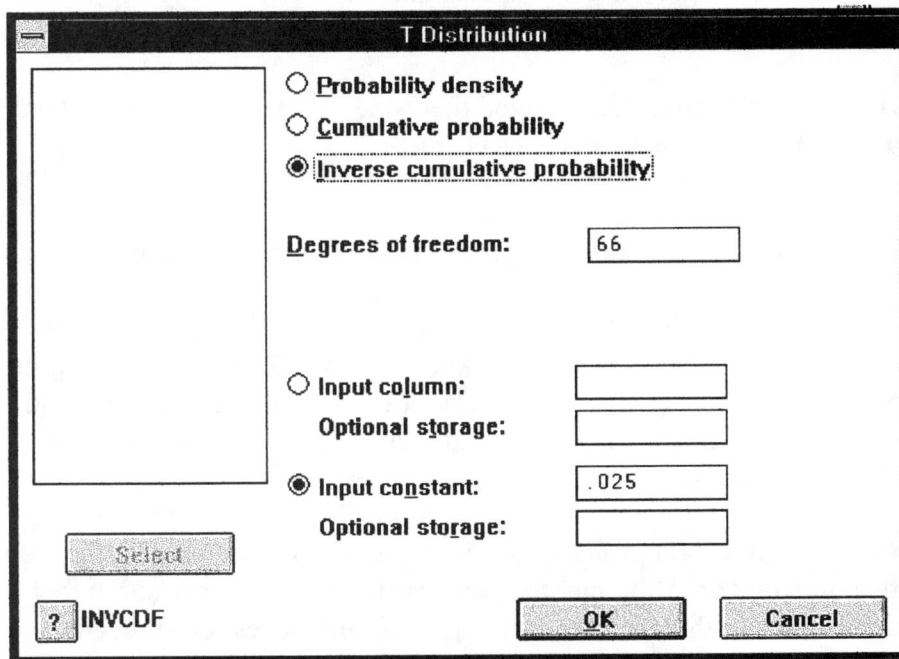

Figure 11.7 T Distribution Dialog Box

After you click on **OK**, the results will appear in the Session window. They are shown in Figure 11.8.

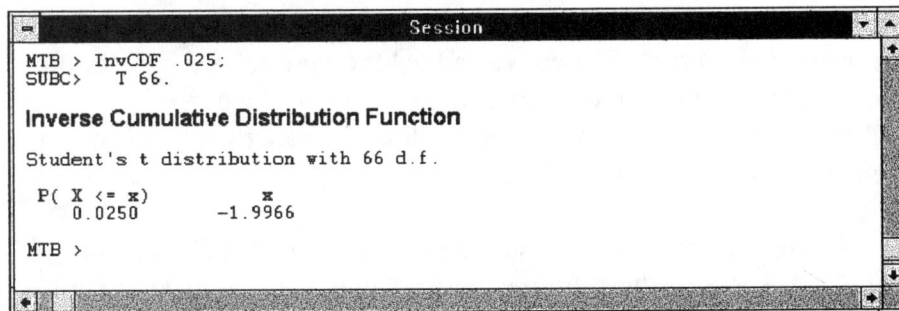

Figure 11.8 Critical Values for a Two-sided T test (df = 66, α = 0.05)

Comparing the test statistic to the critical values -1.99696 and 1.9966 you find that you reject H_O and conclude that mean *Diameter* is significantly different for the two speeds.

You could also reach this decision by looking at the p-value for the test and comparing it to the chosen level of significance. In this case since the p-value is 0.019 is less than the value of α you reject H_o.

Exercise 4. Suppose you changed the value of α for the test from 0.05 to 0.01. How do the conclusions change?

Section 11.4.5 Statistical Differences vs. Meaningful Differences

Although the difference in the mean values of *Diameter* is *statistically* significant you do not know if the difference is meaningful to the manufacturing company. Very often statistical significance and practical significance are not the same thing. The actual value of the difference in the two means is 0.0291 inches.

By now you probably have noticed that the values of *Diameter* are a bit strange. Although it appears that *Diameter* is measured to 4 decimal places, there are not as many *different* values for *Diameter* as you would expect. In fact, there are only 8 different values that the 4 decimal places can be. Can you think of a reason that this might happen?

The real reason is that the value of *Diameter* was measured to the nearest millimeter and converted to inches because the product specifications for the paper towel product are measured in English measures, not metric. The converted values were recorded with four decimal places. Since the original values were to the nearest millimeter, the converted values must differ by exactly 0.0394 (using the conversion factor of 25.4 millimeters per inch).

Thus the statistically significant difference of 0.0291 inches is *smaller* than the smallest difference that the measuring devices could detect! This is an example of how statistically different and practically different are **NOT** necessarily the same. It is also an example of how the accuracy of the measuring device can affect results.

Section 11.5 Investigative Exercises

In the following exercises you are asked to use the skills introduced in the previous chapters to extract information from the tissue strength datafile. You are provided with space to answer the questions and paste in graphical output from the program. If you do not have access to a printer, you can sketch the graphs in the spaces provided.

1. Perform a hypothesis test at the .05 level of significance to determine whether the mean value of *Firmness* differs when the speeds are different.

a) Were you willing to assume that the two population variances were equal?

b) Was there a significant difference in the mean *Firmness* for the two populations?

c) Change the value of alpha to 0.01. How does this change the results of the test? What if you change it to 0.10?

2. Using the procedure outlined in the chapter, set up and perform hypothesis tests to determine whether there is a difference in mean *Diameter* for the two populations defined by *Drawroll*. Do the same thing for *Embosser*. Use a level of significance of 0.01. Fill in the table below:

Population Factor	Sample Sizes	Assume Variances Equal?	Decision Reject H_0/ Fail to Reject H_0
Drawroll			
Embosser			

b) If the differences in the mean *Diameter* are statistically significant, are they of practical significance?

3. Repeat the previous exercise for the variable *Firmness*.

Population Factor	Sample Sizes	Assume Variances Equal?	Decision Reject H_0/ Fail to Reject H_0
Drawroll			
Embosser			

b) Based on these results would you say that there is a significant difference in mean *Firmness* for the two populations defined by *Drawroll*?

c) What about the populations defined by *Embosser*?

4. The towel team is also interested in knowing whether *Diameter* meets the specifications for the roll of towels, 5.35 inches.

a) Perform a hypothesis test to decide whether average *Diameter* for *Speed* = 0 meets the specifications. What test did you use and what is your conclusion?

b) Do the same thing for *Speed* = 1. What test did you use and what is your conclusion?

c) In reality the team is interested in knowing whether the specification is met over the entire range of speeds that are used. Perform a hypothesis test to determine whether *Diameter* meets the specifications for the two populations together. What is your conclusion?

5. Repeat the previous exercise for the variable *Drawroll*.
a)

b)

c)

6. Repeat the exercise for *Embosser*.
a)

b)

c)

7. Prepare a report for management explaining the effects that each of
the machine settings have on *Diameter* and *Firmness*. Indicate which
settings result in acceptable values for roll diameter. Make a
recommendation on machine settings if you can.

Chapter 12 "Are Your Members Happy?"

Testing Proportions

Section 12.1 Overview

Statistical Objectives: After reading this chapter and doing the exercises you will:

- Know how to analyze survey data in order to estimate true population proportions
- Know how to decide whether to use a one-tail test or a two-tail test
- Understand p-values, how they are used and how they relate to alpha
- Know what it means to test the value of p against .50
- Know what the phrase "significantly different" means.

Section 12.2 Problem Statement

Many Americans have taken up golfing. This population of golfers influences several different industries. As we saw in Chapter 5 many golf manufacturers are interested in satisfying their customers by designing golf balls which will fly further. In addition to having an influence on manufacturers, these golfers are important to another sector of the market: country clubs. Many golfers belong to a country club in order to be sure that they can get on the course without a long wait.

In addition to spending time on the golf course, many members also eat meals and socialize at their country club. Thus, if you are on the Board of Directors of such a club, you should be concerned with keeping your members satisfied. One such New England Country Club decided to survey its 250 members. They were interested in member satisfaction with the overall club conditions, the food services and the dues structure. A questionnaire was designed and mailed to all of the club members. Of the 250 current members, 134 returned the survey. This translates to a 53.6% response rate, which is a stronger response than you usually receive from a mail survey.

The data set you will be analyzing in this chapter contains demographic information about the respondent as well as his/her answers to the questions about member satisfaction.

Section 12.3 Characteristics of the Data Set

FILENAME: CHAP12.MTW
SIZE   COLUMNS   15
               ROWS   134

The first 8 columns of the actual datafile are shown in Figure 12.1

	C1	C2	C3	C4	C5	C6	C7	C8
	Howlong	Type	Sex	MStatus	Depends	Age	Income	Area
1	1	1	2	1	0	2	2	2
2	2	2	2	2	2	3	4	1
3	1	1	1	1	0	2	1	2
4	4	2	2		0	4	0	2
5	1	2	1	2	0	4	0	2
6	3	2	2	2	0	3	2	2
7	1	2	1	2	0	4	0	2

Figure 12.1 The first 8 Columns of CHAP12.MTW

The remaining 7 columns are shown in Figure 12.2

	C9	C10	C11	C12	C13	C14	C15	C16
↓	Often	Friend	Condit	Greens	Landscap	Parking	Locker	
1	3	3	2	2	3	1	2	
2	2	4	1	2	2	2	3	
3	1	2	3	3	3	2	2	
4	3	1	3	3	4	3	3	
5	2	1	3	2	3	3	2	
6	3	2	3	3	1	3	3	
7	2	1	3	2	3	3	2	

Figure 12.2 The last 7 Columns of CHAP12.MTW

Notes on the dataset: Non-response is coded zero (0) for all variables.

1. The variable *Howlong* indicates how long the person has been a member of the Country Club:
     1= under 1 year, 2= 1-3 years, 3= 4-7 years,
     4= 8-10 years, 5= more than 10 years.

2. The variable *Type* indicates the type of membership that the member currently has:
     1= Full, 2=Family, 3= Associate, 4= Junior, 5=College,
     6=Social, 7= Restricted Family, 8= Senior Full, 9= Senior
     Associate, 10= Senior Family, 11= Restricted Senior Family.

3. The variable *Sex* indicates the sex of the member:
     1= Male, 2= Female

4. The variable *MStatus* indicates the marital status of the respondent:
     1= Single, 2= Married

5. The variable *Depends* indicates how many dependents the member has.

6. The variable *Age* contains information about the respondents age:
     1= under 20   2= 21-40   3=41-60   4= over 60.

7. The variable *Income* indicates the annual income range of the member:

>   1= under $25,000   2= $25,001 - 50,000
>   3= $50,001-75,000   4= $75,001 or more.

8. The variable *Area* tells if the member has ever belonged to any other country club in the area: 1= Yes, 2 = No.

9. The variable *Often* indicates how often the respondent uses the golf course:

>   1= less than once a week,   2= 1-2 times a week,
>   3= 3-4 times a week, 4= more than 4 times a week.

10. The variable *Friend* contains information regarding how often the member brings a guest to play golf at the club:

>   1= Always, 2= Frequently, 3= Sometimes, 4= Never.

The last 5 variables indicate the respondents ranking of 5 features of the club. The following numeric rating scheme is used:

>   1=Poor, 2= Fair, 3= Good, 4= Excellent.

11. The variable *Condit* is the respondents rating of the condition of the golf course.

12. The variable *Greens* is the respondents rating to of the condition of the greens.

13. The variable *Landscap* is the respondents rating of the landscape surrounding the golf course.

14. The variable *Parking* is the respondents rating of the accessibility of parking.

15. The variable *Locker* is the respondents rating of the conditions of the locker rooms.

Use File>Open Worksheet to read in the file.

Read in the datafile named CHAP12.MTW using the commands described in Chapter 3. Remember to create a working version of the datafile by resaving the file with a slightly different name.

Section 12.4 Creating Cross Tabulation Tables in Minitab

In order to perform a hypothesis tests on proportions using Minitab, you must provide the program with sample proportions. For example, the sample proportion of interest might be the proportion of respondents who ranked the condition of the greens as "Excellent". This would be calculated by finding the number of respondents who ranked the condition of the greens as "excellent" and dividing by the total number of respondents.

If your data set contains raw survey data like the one in this chapter, then you must first tabulate the data in order to obtain the sample proportion. Thus, before you learn how to run hypothesis tests on proportions you must learn how to create the appropriate tables. From the menu bar click on **Stat**. The pull-down menu shown in Figure 12.3 will appear

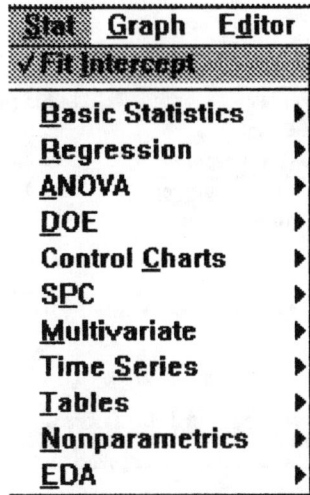

Stat   Graph   Editor
√ Fit Intercept

Basic Statistics   ▶
Regression   ▶
ANOVA   ▶
DOE   ▶
Control Charts   ▶
SPC   ▶
Multivariate   ▶
Time Series   ▶
Tables   ▶
Nonparametrics   ▶
EDA   ▶

Figure 12.3 The Stat Menu

From this menu choose **Tables** and the pop-up menu shown in Figure 12.4 will appear.

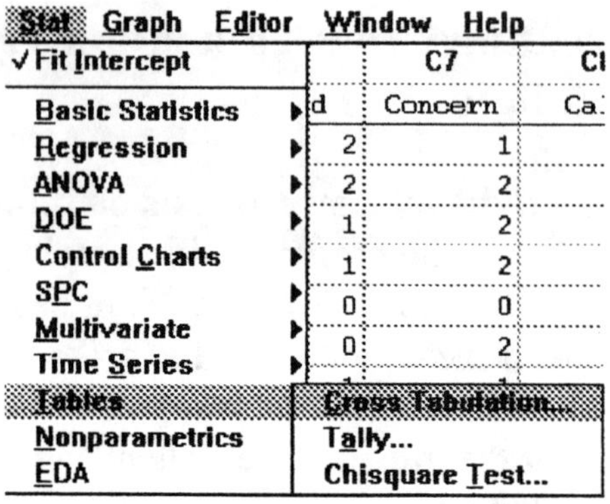

Figure 12.4 Tables Pop-up Menu

From this screen select the **Cross_Tabulation** option and then you will see the Cross Tabulation Dialog box shown in Figure 12.5.

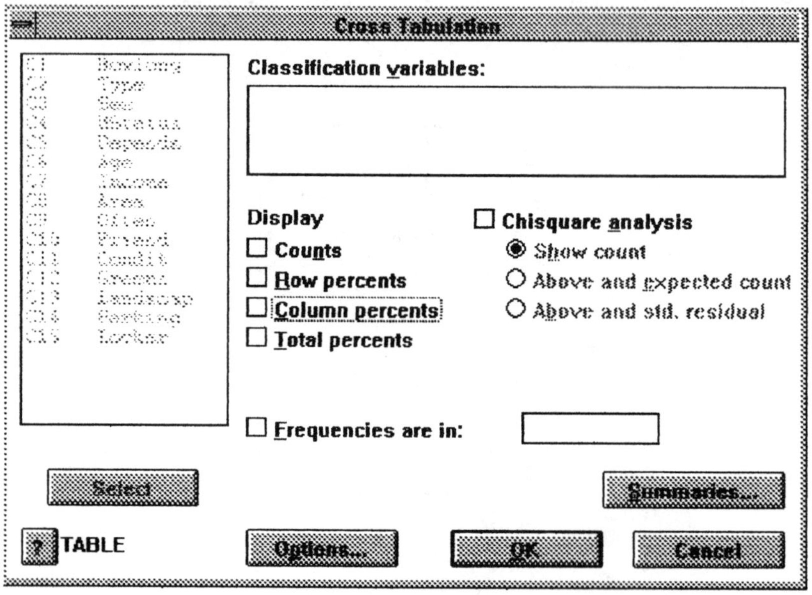

Figure 12.5 Cross Tabulation Dialog Box

Position the cursor in the **Classification variables** box and then select two variables from the list at the left of the box. For example, double click on the variable *Condition* and then double click on the variable *Sex*. In order to get the most complete information from Minitab regarding the simultaneous behavior of the variables *Condition* and *Sex*, in the Display area portion of the Dialog box, click on the boxes called **Counts**, **Row percents** and **Column percents**. Now click **OK** and you will see the cross tabulation table shown in Figure 12.6 displayed in the Session window.

Remember to save your Session window if you wish to save the cross tabulation tables.

```
ROWS: Condit      COLUMNS: Sex

            0         1         2       ALL

0           0         0         2         2
           --        --    100.00    100.00
           --        --      7.14      1.49

1           0         0         1         1
           --        --    100.00    100.00
           --        --      3.57      0.75

2           1        26         8        35
         2.86     74.29     22.86    100.00
       100.00     24.76     28.57     26.12

3           0        67        15        82
           --     81.71     18.29    100.00
           --     63.81     53.57     61.19

4           0        12         2        14
           --     85.71     14.29    100.00
           --     11.43      7.14     10.45

ALL         1       105        28       134
         0.75     78.36     20.90    100.00
       100.00    100.00    100.00    100.00

CELL CONTENTS --
              COUNT
              % OF ROW
              % OF COL
```

**Figure 12.6 Cross Tabulation of *Condition* by
*Sex***

As you examine the table, you see the possible values for the variable *Condition* (0,1,2,3,4) listed down the left hand side of the table and the possible values for the variable *Sex* (0,1,2) listed across the top of the table. There is also a row and column labeled ALL providing the row and column totals respectively. In this example, the variable *Condition* is referred to as the row variable and *Sex* is referred to as the column variable. Remember that in specifying the variables in the Dialog box, we selected the variable *Condition* first. Thus the first variable selected becomes the row variable.

Each cell in the table contains 3 values: (1) a frequency, (2) that frequency as a percent of the row total and (3) that frequency as a percent of the column total. If you look at the bottom of the table shown in Figure 12.6 you will see a key detailing the contents of each cell. For example, there were 15 female respondents (column variable coded #2) that ranked the condition of the golf course as good (row variable coded #3).   This corresponds to 18.29% of the respondents who ranked the condition of the golf course as good and 53.57% of all the responses of the women on this issue.

Exercise 1. How many male respondents ranked the condition of the golf course as good? What percentage is this of the respondents who ranked the condition of the golf course as good? What percentage is this of the male respondents?

Exercise 2. If you were only interested in the variable *Condition*, what other Minitab command sequence could you have used to obtain the frequency distribution for just the single variable?   Hint: you learned it in one of the early chapters.

Section 12.5 Hypothesis Test on a Single Proportion Using Minitab

The next step is to use Minitab to run a hypothesis test on a single proportion.   Unfortunately Minitab does not have a built in test for proportions as it does for means. It does, however, have a built in feature to standardize any value by subtracting a mean and dividing by a standard deviation. This is precisely what we need since the test statistic for a test on a single proportion is a Z statistic :

$$Z = \frac{\hat{p} - p}{\sqrt{p(1-p)/n}}$$

where: $\hat{p}$ is the sample proportion from the data

   p   is the value you are testing it against

and   n   is the sample size.

The denominator of this expression is the standard error of the estimate, $\hat{p}$. In order to get Minitab to calculate the Z test statistic, you need to specify three values: (1) the sample proportion which you would read from the frequency table or the cross tabs which you have created; (2) the hypothesized proportion; and (3) the sample size. Let's continue to examine the variable *Condition*.

Suppose the Board of Directors wishes to advertise that more than half of their members feel that the condition of the golf course is "good". From the cross tabulation table shown in Figure 12.6 you can see that 82 members rated the condition of the golf course as "good" (coded 3). This is the first step in obtaining the sample proportion to be used in the hypothesis test. The sample proportion would be 82/134 = .6119 if there were no non-response.

Section 12.5.1 Handling the Item Non-Response

Notice that there were two respondents who did not rank the condition of the greens. These are coded zero (0). Because of this the proportions should be calculated out of 132 responses instead of out of 134 responses to this question. In order to avoid this problem, it is best to first sort the variable you are interested in studying. Then when you create the frequency table or cross tabulations, you would leave the zeros out of the tabulation. If they are all at the top or the bottom of the column this is easy to do. Alternatively, you can use the information from the cross tabulation table you have already created and simple calculate the sample proportion yourself by reducing the sample size by the number of non-responses for that questions.

Sort the data first in order to handle the item non-response.

Continuing with this example, the sample proportion is found to be
82/132 = .6212.

Section 12.5.2 Using Minitab to find the Test Statistic

Type the value for the sample proportion into the first cell in an
empty column in the spreadsheet. In this case, enter the value of .6212
in the first row of column C16. From the menu bar select **Calc** and you
will see the menu options shown in Figure 12.7.

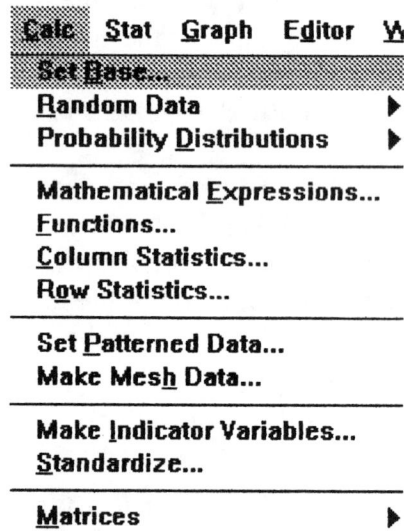

Calc   **Stat**   **Graph**   **Editor**   **W**
Set Base...
Random Data   ▶
Probability Distributions   ▶
Mathematical Expressions...
Functions...
Column Statistics...
Row Statistics...
Set Patterned Data...
Make Mesh Data...
Make Indicator Variables...
Standardize...
Matrices   ▶

Figure 12.7 The Calc Menu

From this menu select **Standardize** and you will see a dialog box
similar to the one shown in Figure 12.8

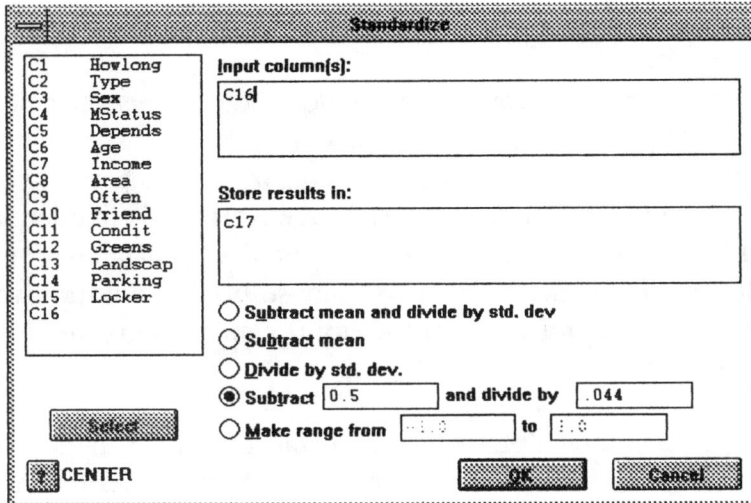

**Figure 12.8 The Completed Standardize Dialog
Box**

Notice that the dialog box shown in Figure 12.8 has already
been completed. You must specify (1) the **Input column**, (2) the
location to **Store results in**, and (3) click the circle next to the words
Subtract _____ and divide by _____. The **Input column** is the
column number which contains the sample proportion, $\hat{p}$. In our
example we have placed $\hat{p}$ in column C16. Column C17 is empty and
can be used for the results. We wish to subtract the value of the
proportion we are testing, in this case the hypothesized proportion of
.50 is the true mean proportion we are testing against. We wish to
divide by the standard error which is $\sqrt{p(1-p)/n} = \sqrt{.5(1-.5)/132} =$
.044. When all of these values have been entered the dialog box will
look like the one shown in Figure 12.8. Click **OK** and you will find
the Z test statistic placed in column C17. For this example the Z test
statistic is 2.75455.

Section 12.5.3 Deciding to Fail to Reject H_0 or to Reject H_0

Use the critical Z value(s) to decide if you should reject the null hypothesis.

There are two ways to decide if you should reject the null hypothesis (H_0). One way is to compare the Z test statistic to the critical Z value(s) for a given value of alpha, α. In Chapter 10 you learned you to find the critical values for the Normal distribution using the **Calc>Probability Distributions>Normal** command in Minitab. In the dialog box, remember that you choose **Inverse cumulative probability** and enter a value for the **Input Constant** in order to get the critical Z value.

There are three different cases which you may encounter when doing a hypothesis test on proportions. The critical values for each case are found using the approach you learned in Chapter 10. The only difference is the value which you supply as the **Input Constant.**

The first two cases are shown below. In these cases you are doing a one sided test on proportions. The specific value of p being tested is p=.50.

Case I: H_O: $p \geq .50$ Case II: H_O: $p \leq .50$
                H_A: $p < .50$ H_A: $p > .50$

For Case I, you would reject H_0 if the Z statistic is "too small" or less than the negative one-sided critical value. This critical value is found by entering the value of alpha being used for the test as the **Input constant**. If alpha is .05 you should verify that the critical value is -1.6449.

Reject H_0 if the test statistic falls beyond the critical values.

For Case II, you would reject H_0 if the Z statistic is "too large" or greater than the positive one-sided critical value. This critical value is the same as that found for Case I only with the sign changed to a positive number.

The third case is a two sided test on proportions. A two sided test against the true value of p=.50 is shown below.

Case III: H_0: $p = .50$
                    H_A: $p \neq .50$

For Case III, you would reject H_0 if the Z statistic is smaller than the negative two-sided critical value or larger than the positive two-sided critical value. The negative two-sided critical value is obtained by entering $\alpha/2$ as the **Input Constant**. The positive two-sided critical value is obtained by changing the sign. You should verify that the lower and upper critical values are -1.96 and 1.96 when $\alpha = .05$.

To determine whether you should use a one-sided test or a two-sided test you need to consider what question you are trying to answer. For the example if management would like to say that 50% or more of their members ranked the condition of the greens as "good" then you would use Case I of the one-sided test and since the Z-statistic is 2.75455 , you would reject the null hypothesis.

You can alter the critical regions by changing the value for alpha from .05. Thus, your decision to reject the null hypothesis could be different for different values of alpha.

Exercise 3: Change alpha to .01. What happens to the critical values? Do you still reject H_0? Now change alpha to .10. What happens to the critical values? Do you still reject H_0?

Use the p-value to decide if you should reject the null hypothesis.

The second method of deciding whether to reject or fail to reject H_0 is to use the p-value. Remember that the p-value for a test of an hypothesis is the probability of obtaining a value of Z as extreme or more extreme than the sample value when H_0 is true. The p-value is also called the observed significance level of the test since it represents the smallest value of alpha for which we could reject H_0 using the observed data. For example, suppose the p-value was .07. This means that if you set alpha at .05 you would fail to reject H_0 but if you set alpha at .10 then you would reject H_0. In fact the decision switches at $\alpha = .07$. If the p-value is smaller than alpha then you reject H_0. By providing management with the p-value they can then see for what value of alpha the decision changes.

*Remember the **p-value** tells you at what value of alpha the decision switches.*

The p-value can be found by using the Cumulative distribution for the Normal distribution which you learned about in Chapter 7. The Minitab command sequence is **Calc>Probability Distributions>Normal**. In the dialog box you choose **Cumulative distribution** and enter the Z test statistic for the **Input Constant**. For our example the Z test statistic is 2.75455. The probability which is calculated and shown in the Session window is labeled P(X<=x). For our example, this probability is .9971. When the Z test statistic is positive, the p-value is calculated by subtracting this probability from 1. So our p-value in this case is 1-.9971=.0029. When the Z test statistic is negative, the p-value is precisely the cumulative probability given by Minitab.

One of the important things to remember when conducting any hypothesis test is that as you decrease the probability of a Type I error, α, the probability of a Type II error, β, increases. Thus, it is not always desirable to set α as small as possible because in doing so the probability of a Type II error increases dramatically. By reporting the p-value , the decision to reject the null hypothesis (with the potential for a type I or a type II error) is left up to the decision maker.

Exercise 4:   What is the p-value for the one-sided test hypothesis test on the variable *Condition*? Is it legitimate for the Board to make the statement they wish regarding how their members feel about the condition of the course?

Section 12.6 Investigative Exercises

1. Describe the respondents (demographics) of the survey of the Country Club members. Use whatever graphs and descriptive statistics you feel are appropriate.

2. The response rate for this survey was 53.6% which is high for a mail survey. Even though there was a high response rate, the results could still be biased by the non-response. Bias creeps in if the non-respondents "look" different from the respondents. You should try to get some of the non-respondents to respond by using a second mailing or a phone interview. Speculate about the type of member who would be inclined to not respond. What would you recommend to the Board of Directors.

3. Complete the following table with the appropriate proportions:

Rated the:	Exc.	Good	Fair	Poor
Condition of the Golf course				
Condition of the greens				
Condition of the landscape				
Parking accessibility				
Condition of the locker rooms				

What areas seem to be in need of improvement?

General Instructions for Exercises 4-7: No alpha value is specified for the hypothesis tests in exercises 4-7. You should report the p-values and explain what the p-value means in each case.

4. Examine the proportion of respondents who rated the *condition* of the greens "good". Conduct the following hypothesis test:

$$H_0: \quad p \geq .50$$
$$H_A: \quad p < .50$$

What is your conclusion about the true proportion of members who feel the condition of the greens is "good" ?

5. What proportion of the respondents rated the *landscape* "fair" or "poor"? Does this mean that you would automatically reject the null hypothesis in the following test:

$$H_0: \quad p \leq .20$$
$$H_A: \quad p > .20$$

Conduct the hypothesis test to verify your answer.

6. Can you conclude that less than 5% feel that the *parking* accessibility is "poor"?

7. Can you conclude that more than 50% feel that the *condition of the locker rooms* is "poor"?

8. On the basis of your analysis, what areas do you feel need improvement? What recommendations do you have for increasing membership for this Country Club?

Chapter 13

"Country Club Members: A Closer Look"

The Chi Square Distribution

Section 13.1 Overview

Statistical Objectives: After reading this chapter and doing the exercises you will:

- Know how to use a Chi Square test to compare two proportions
- Know when it is appropriate to use a Chi Square test for independence
- Know how to interpret the results of a test for independence.

Section 13.2 Problem Statement

In Chapter 12 you looked at the results of a survey of the members of a New England country club. In particular, you examined a variety of questions related to member satisfaction. In that chapter you examined single proportions. You were able to draw some conclusions based on this analysis. In this chapter you will continue to analyze this survey data.   For example, you may wish to know whether members in two or more age groups have the same opinion about the condition of the golf course. You may also wish to know if a member's income is related to his/her rating of the landscape surrounding the golf course.

Read in the datafile named CHAP12.MTW using the commands describe in Chapter 3. Remember to create a working version of the datafile by resaving the file with a slightly different name.

Section 13.3 Comparing Two Proportions: Hypothesized Difference of Zero

Section 13.3.1 Setting up the Hypothesis Test

In many cases it is desirable not only to look at a single proportion but to compare proportions. For example, the Board of Directors may wish to compare the responses of the young members to the older members. They may wish to compare the responses of the newer members to the members who have been with the Club many years. Another possibility would be to compare the responses of the male members to the female members. Perhaps the married members have different concerns than the single members. All of these comparisons can be examined by running a hypothesis test to compare two proportions.

In this section you will examine whether the locker rooms are rated the same by the men and the women. In particular we will compare the proportion of males who rated the locker rooms "good" to the proportion of females who rated the locker rooms "good". Call the male respondents population 1 and the female respondents population 2. The proportions calculated for population 1 are then labeled p_1 and those from population 2 are labeled p_2. The comparison hypothesis test can take one of three forms:

Case I: $H_O : p_1 - p_2 \leq 0$ **Case II:** $H_O : p_1 - p_2 \geq 0$
$\quad\quad\quad H_A : p_1 - p_2 > 0$ $\quad\quad\quad\quad\quad\quad\quad H_A : p_1 - p_2 < 0$

Case III: $H_O : p_1 - p_2 = 0$
$\quad\quad\quad\quad H_A : p_1 - p_2 \neq 0$

In order to decide which of these three tests you should use, you need to consider what type of information you desire. Most of the time Case III is used to examine whether or not the two groups behave *differently*. For the example use Case III to see if the men rate the locker rooms differently from the women.

Section 13.3.2 Comparing Two Population Proportions in Minitab

You most likely learned that the appropriate test statistic for comparing two population proportions is a Z test statistic which looks like this:

$$z = \frac{(p_1 - p_2) - 0}{\sqrt{\dfrac{\bar{p}(1-\bar{p})}{n_1} + \dfrac{\bar{p}(1-\bar{p})}{n_2}}}$$

where p_1 and p_2 are the sample proportions and $\bar{p}$ is the pooled proportion found by $\dfrac{n_1' + n_2'}{n_1 + n_2}$. The values for n_1' and n_2' are the number in the first and second sample, respectively, with the characteristic of interest. For the example, that is the number of men who rated the condition of the locker rooms as "good" (n_1') and the number of women who rated the condition of the locker rooms as "good" (n_2'). The pooled proportion is an estimate of the true proportion of members who would rate the locker rooms as excellent if H_0 is true.

Unfortunately Minitab does not provide a test of two proportions using the Z test statistic. However, the extension of a test comparing two proportions is to compare more than two proportions. This is done using a Chi Square test. Although not traditionally presented in this fashion, there is no reason why a test on two proportions can not be done using the more generalized test, the Chi Square test. This is what we will do in order to get Minitab to do the calculations for us.

Use the Chi Square test to compare 2 proportions in Minitab.

The Chi Square test statistic can be calculated in two different commands in Minitab. The **Stat>Tables>Cross Tabulation** command sequence which you learned to use in Chapter 12, has an option to have Minitab calculate a Chi Square statistic for you. This should be used if you have raw data. The other way is to use the **Stat>Tables>Chisquare Test** command sequence. This can only be used if your data is stored in the spreadsheet as frequencies. If you have raw data, as in the case of the Country Club, then you must first create the frequencies using the **Cross Tabulation** option.

Use Stat>Tables>Cross Tabulation if you have raw data.

Fortunately Minitab allows you to specify the calculation of the Chi Square test statistic at the same time.

Section 13.3.3   Converting Multinomial data to Binomial Data in Minitab

Since we are only interested in comparing the proportion of men and women who rated the condition of the locker rooms as "good", we must first recode the data so that we have binomial data instead of multinomial data. Binomial data (remember from Chapter 7) is data that has only two possible responses. Right now the variable *Locker* has 4 possible responses (1-4). Multinomial data is data that has more than 2 possible responses. We only want to consider the "good" ratings right at the moment. The natural extension will be to consider all 4 of the possible responses. We will do this in Section 13.4.

The first step is then to use the recoding ability of Minitab to reduce the ratings to only 0's and 1's. We wish to recode all of the "good" ratings (coded 3) to be 1's and all other responses to be 0's. From the main menu bar choose **Manip** and you will see the menu shown in Figure 13.1

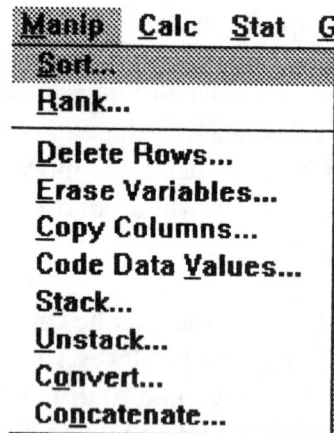

Figure 13.1 The Manip Menu

From this menu choose **Code Data Values** and you will see a dialog box similar to the one shown in Figure 13.2.

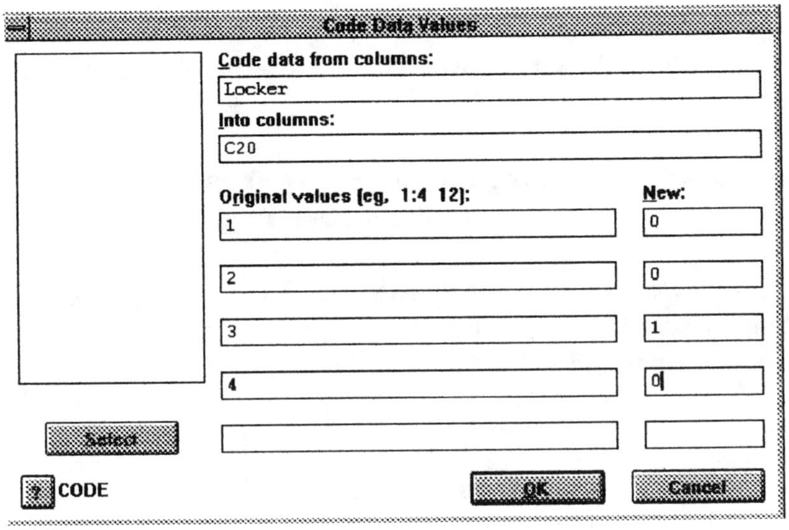

Figure 13.2 Completed Dialog Box to Recode
Locker **Variable**

The dialog box shown in Figure 13.2 is completed. Notice that the variable named *Locker* is the data to be coded and it will be place into the empty column C20. The original values and the new values are then entered and when all of the information has been provided click on **OK.**

A portion of the new (binomial) version of the variable *Locker* and the original data are shown in Figure 13.3 and you can see that the task has been accomplished!

C15	C20
Locker	
	0
3	1
2	0
3	1
2	0
3	1
2	0
3	1

Figure 13.3 Original data and recoded data for
Locker **variable**

Exercise 1. Recode the variable *Greens* to allow for the comparison of the proportion of men and women who feel that the condition of the greens is "poor".

Section 13.3.4 Calculating the Chi Square Test Statistic to Compare Two Population Proportions

Now we are ready to calculate the Chi Square test statistic to decide if we should reject or fail to reject the null hypothesis. Recall that we are testing the following hypothesis:

$$H_O: p_1\text{-}p_2 = 0$$
$$H_A: \ p_1\text{-}p_2 \neq 0$$

where: p_1 is the proportion of males who rated the conditions of the *Locker* "good"

and p_2 is the proportion of females who rated the conditions of the *Locker* "good".

From the main menu bar choose **Stat>Tables>Cross Tabulation** and complete the dialog box so that it looks like the one shown in Figure 13.4.

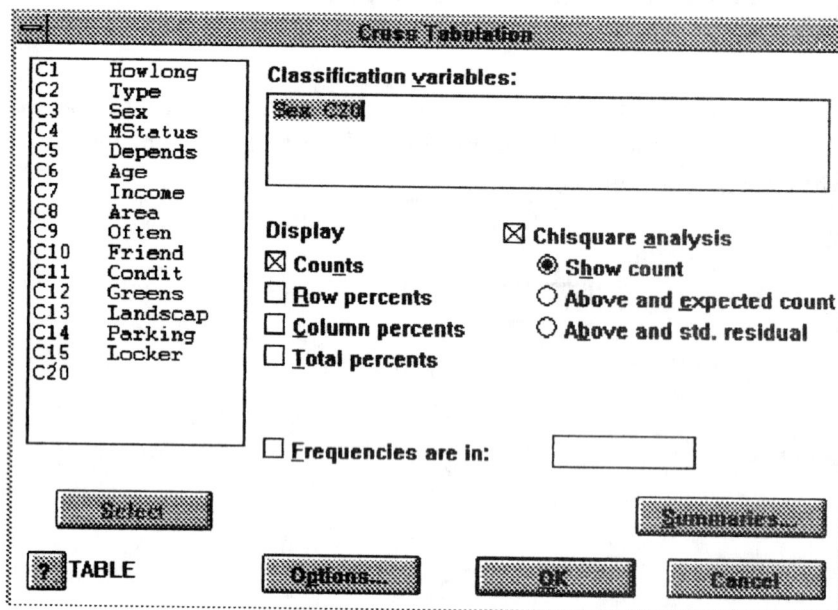

Figure 13.4 Completed Dialog Box to Generate Cross Tabs and the Chi Square Statistic

Be sure the box next to the words **Chisquare _a_nalysis** is checked. Click **O_K_** when your dialog box has all the information. You should see the cross tabulation table and the Chi-Square test statistic in the Session window as shown in Figure 13. 5.

```
ROWS: Sex      COLUMNS: C20

                 0          1        ALL

    0            1          0          1
                 1          0          1

    1           76         29        105
                76         29        105

    2            9         19         28
                 9         19         28

  ALL           86         48        134
                86         48        134

CHI-SQUARE  =       16.131   WITH D.F. =      2

    CELL CONTENTS --
                       COUNT
                       COUNT
```

Figure 13.5 Cross Tabulation Table of *Sex* by
Locker (recoded)

Observe that the Chi Square Test statistic is 16.131 with 2 degrees of freedom (D.F.) Based on this test statistic value you must decide to reject or fail to reject H_o. This is covered in the next section.

Exercise 2. Find the Chi Square test statistic to test to see if there is any difference in the proportion of men and women who rated the condition of the *Greens* "poor".

Section 13.3.5 Deciding to Fail to Reject H_0 or to Reject H_0

As with any hypothesis test, you may use either the critical value or the p-value to decide whether to reject H_0. In this case the critical value is a Chisquare critical value. Unlike the Z and the t distributions, the Chisquare distribution is not symmetric. This means you can not just change the sign to get the upper critical value from the lower critical value.

*Use **Calc>Probability Distribution>Chisquare** to get the critical values. Be sure to click on the Inverse cumulative probability button.*

For Case I, you would reject H_0 if the Chisquare statistic is "too small" or less than the lower critical value. This lower critical value is found by entering the value of alpha being used for the test as the **Input constant**. If alpha is .05 and there are 2 degrees of freedom you should verify that the critical Chisquare value is .1026.

Reject H_o if the test statistic falls beyond the critical values.

For Case II, you would reject H_O if the Chisquare statistic is "too large" or greater than the upper critical value. This upper critical value is found by entering the value of (1-alpha) as the **Input constant**. If alpha is .05, you should enter .95 as the **Input constant**. Verify that the upper critical Chisquare value with 2 degrees of freedom is 5.9915

For Case III, you would reject H_O if the Chisquare statistic is smaller than the lower critical value or larger than the upper critical value. The lower critical value is obtained by entering $\alpha/2$ as the **Input Constant**. The upper critical value is obtained by entering $(1-\alpha/2)$ as the **Input Constant**. You should verify that the lower and upper critical values are .0506 and 7.3778 when $\alpha=.05$ and there are 2 degrees of freedom.

Exercise 3. Using the lower and upper critical values found above, what can you conclude about the proportion of men and women who feel that the locker rooms are in "good" condition? What suggestions might you have for the Board of Directors?

Exercise 4. Using the lower and upper critical values found above, what can you conclude about the proportion of men and women who feel that the condition of the green is "poor"?

Use the p value to decide if you should reject H_o

The second method of deciding whether to reject or fail to reject H_O is to use the p-value. Remember that the p-value for a test of an hypothesis is the probability of obtaining a value of Chisquare as extreme or more extreme than the sample value when H_O is true. The p-value is also called the observed significance level of the test since it represents the smallest value of alpha for which we could reject H_O using the observed data. For example, suppose the p-value was .07. This means that if you set alpha at .05 you would fail to reject H_O but if you set alpha at .10 then you would reject H_O. In fact the decision switches at $\alpha = .07$. If the p-value is smaller than alpha then you reject H_O. By providing management with the p-value they can then see for what value of alpha the decision changes.

The p-value can be found by using the Cumulative distribution for the Chisquare distribution. The Minitab command sequence is **Calc>Probability Distributions>Chisquare**. In the dialog box you choose **Cumulative distribution** and enter the Chisquare test statistic for the **Input Constant**. For our example the Chisquare test statistic is 16.131. The probability which is calculated and shown in the Session window is labeled $P(X<=x)$. For our example, this probability is .9997. The p-value is calculated by subtracting this probability from 1. So our p-value in this case is 1-.9997 =.0003.

*Remember the **p-value** tells you at what value of alpha the decision switches.*

Section 13.4 Chi-Square Test for Independence

In order to extend the test on two proportions to compare all four of the ratings of the variable *Locker* by men and women, you need to use a Chi Square test for independence. Two criteria of classification are said to be *independent* if the distribution of one criterion in no way depends on the distribution of the other. For example, consider the variables *Age* and *Condit*. You wish to be able to tell the Board of Directors if the different age groupings rated the condition of the golf course the same way. If they are not the same then at least one of the age groupings may have some special needs or concerns which are not being addressed.

Look at the cross tabulation table of *Age* by *Condit* shown in Figure 13.6

```
ROWS: Age      COLUMNS: Condit

          0        1        2        3        4       ALL

  0       0        0        1        1        1        3
          0        0        1        1        1        3

  1       0        0        2        1        1        4
          0        0        2        1        1        4

  2       0        0        8        9        0       17
          0        0        8        9        0       17

  3       0        1        7       32        8       48
          0        1        7       32        8       48

  4       2        0       17       39        4       62
          2        0       17       39        4       62

 ALL      2        1       35       82       14      134
          2        1       35       82       14      134

CHI-SQUARE =     18.893    WITH D.F. =    16
```

Figure 13.6 Observed Frequencies of *Age* by
Condit(ion)

You can see that 1 of the members under 20 (row variable coded 1) rated the condition of the golf course as "excellent" (column variable coded 4), 1 rated it "good", 2 rated it "fair" and 0 rated it "poor". The question of interest is whether or not the distributions of the ratings are the same for the members aged 21-40, 41-60 and over 60. If, in fact, it appears that the age groups rated the condition of the golf course the same way then the variables *Condit* and *Age* are considered independent.

The hypothesis you are testing is:

H₀: *Age* and Rating of the *Condition* of the Golf Course are independent

Hₐ: *Age* and Rating of the *Condition* of the Golf Course are not independent

The following steps should be followed in a Chi Square test of independence.

1. From the population of interest, draw a simple random sample.

2. Display the data in a cross tabulation table. The levels of one criterion provide row headings and the levels of the other criterion provide column headings. The cell entries in this table are called observed frequencies. Such a table is also called a contingency table.

3. Compute the χ^2 statistic and compare it to the upper critical value.

4. If the computed χ^2 statistic is equal to or greater than the upper critical χ^2, then reject H_O and conclude that the variables are not independent. Otherwise do not reject H_O and conclude that the variables are independent.

*Remember that if 2 variables are **independent** it means that **how observations are categorized on one variable has no bearing on how they are categorized on the other variable.***

The first step, data collection, has been done for you. As we have seen in Section 13.3, Minitab calculates the cross tabulation table and the χ^2 statistic for us. Thus steps 2 and 3 are done for us by Minitab. Finally, we learned how to find critical χ^2 values in Section 13.3.5. There is no additional Minitab commands we need then to complete this test for independence.

By examining the output shown in Figure 13.6 you can see that the Chi Square statistic is 18.893. You would reject H_O if the Chi Square statistic was larger than the upper critical value which is 26.2963. Thus, we would fail to reject H_O since the calculated test statistic value is not greater than the critical value. We conclude that *Age* and *Condition* are independent variables.

*Use **Calc>Probability Distributions>Chisquare** and choose inverse cumulative probability to get the critical χ^2 value.*

You should remember that one of the conditions for using the Chi Square test is that there be a minim of 5 observations in each cell of the contingency table. This condition is violated in our example of *Age* and *Condition*. This means that there were too few (<5) observations in some of the cells and you should collapse some of the categories and re-run the test. In order to do this you may have to combine two or more rows or columns together in the observed frequencies. This can be done by recoding the data as we did in the case of two proportions.

Section 13.5 Investigative Exercises

General Instructions for Exercises 1-8: No alpha value is specified for the hypothesis tests in exercises 1-8. You should report the p-values and explain what the p-value means in each case.

1. Is there any significant difference in the rating of the *condition* of the golf course by sex? by age? by income? by marital status? Examine only the "fair" rating. What are your conclusions about the members who consider the condition of the golf course "fair"? Consider using the following table to display your results:

Sex	Proportion rated 'fair'	*Age*	Proportion rated 'fair'
Male		40 or Under	
Female		Over 40	
Significantly Different?		Significantly Different?	

Income	Proportion rated 'fair'	*Marital Status*	Proportion rated 'fair'
Under $50,000		Single	
Over $50,000		Married	
Significantly Different?		Significantly Different?	

2. Is there any significant difference in the rating of the condition of the *greens* by sex? by age? by income? by marital status? Examine only the "fair" rating. What are your conclusions about the members who consider the condition of the greens "fair"?

Sex	Proportion rated 'fair'	Age	Proportion rated 'fair'
Male		40 or Under	
Female		Over 40	
Significantly Different?		Significantly Different?	

Income	Proportion rated 'fair'	Marital Status	Proportion rated 'fair'
Under $50,000		Single	
Over $50,000		Married	
Significantly Different?		Significantly Different?	

3. Is there any significant difference in the rating of the surrounding *landscape* of the golf course by sex? by age? by income? by marital status? Examine only the "fair" rating. What are your conclusions about the members who consider the landscape "fair"?

Sex	Proportion rated 'fair'	Age	Proportion rated 'fair'
Male		40 or Under	
Female		Over 40	
Significantly Different?		Significantly Different?	

Income	Proportion rated 'fair'	Marital Status	Proportion rated 'fair'
Under $50,000		Single	
Over $50,000		Married	
Significantly Different?		Significantly Different?	

4. Is there any significant difference in the rating of the *parking* accessibility by sex? by age? by income? by marital status? Examine only the "fair" rating. What are your conclusions about the members who consider the parking accessibility "fair"?

Sex	Proportion rated 'fair'	Age	Proportion rated 'fair'
Male		40 or Under	
Female		Over 40	
Significantly Different?		Significantly Different?	

Income	Proportion rated 'fair'	Marital Status	Proportion rated 'fair'
Under $50,000		Single	
Over $50,000		Married	
Significantly Different?		Significantly Different?	

5. Investigate the members rankings of the
   condition of the golf course
   condition of the *greens*
   condition of the *landscape* surrounding the golf course
   parking accessibility
   condition of the *locker* room.

Are any of these variables dependent on the members age, sex, gender or income? What recommendations do you have based on this information.

6. Is the variable *Howlong* dependent on the variable *Often*? What does your analysis tell you about the membership at this country club?

7. Is the variable *Type* related to the members age ? Is this surprising?

8. Conduct tests for independence on any other variables that you feel should be examined.

9. On the basis of your analysis, what recommendations would you make to the Board of Directors of this country club. Your recommendations should identify any particular groups who might have special needs and/or concerns, suggest ways to increase their membership and identify areas in need of improvement.

Chapter 14 "How Are They Related?"

Linear Regression and Correlation

Section 14.1 Overview

Statistical Objectives: After reading this chapter and doing the exercises a student will:

- Know the difference between a dependent and independent variable and how to determine which is which.
- Know how to find a linear regression equation for a pair of variables.
- Know how to interpret the output of a linear regression to determine whether a significant relationship exists.
- Know what the coefficient of determination is and what it means.
- Know how to use the regression equation to predict values of the dependent variable.
- Know what residuals are and what they mean.
- Know the difference between interpolation and extrapolation.

Section 14.2 Problem Statement

A company that manufactures computer storage media is wondering whether the **Total Quality Management (TQM)** that they have introduced in their floppy diskette manufacturing line is achieving the results they expected.

The program, which incorporates among other things, Quality Circles, Statistical Quality Control and Team Based Decision Making has been in place for almost two years. According to the literature when such programs are used they should result in increased productivity and quality and decreased waste and delay. The management of the company asked the production team to assemble some data to assess the success of the program before they introduce similar programs into other areas of the corporations.

Before they collected any data, the team assembled a checklist of what they think they know about the process and its behavior over the past two years. They all agree that they have noticed an increase in the speed at which the production line runs and most people thing that there has been a decrease in the amount of waste from the process. They are not sure that the increase in speed means that there has been an increase in productivity, nor are they really sure that there has been a decrease in delay. They are almost all sure that they saw an initial improvement in quality, but that the improvement was not sustained.

They decided to look at five different variables from the production process to see if their initial reactions to the question are correct. They then collect monthly data on Machine Speed, Waste, Delay, Rate of Operation and Average Outgoing Quality for the two year period since the TQM program was introduced.

Section 14.3 Characteristics of the Data Set

FILENAME: **CHAP14.MTW**
SIZE: **COLUMN   6**
                 ROWS   25

The first seven lines of the datafile are shown in Figure 14.1.

	C1	C2	C3	C4	C5	C6	C7
	MONTH	SPEED	WASTE	RATEOPER	DELAY	QUALITY	
1	1	375	8.9	14.5	6.26423	95	
2	2	334	9.9	12.8	6.48542	93	
3	3	356	8.7	12.7	6.83717	94	
4	4	378	9.5	13.9	5.71337	93	
5	5	373	9.8	13.7	6.31361	94	
6	6	381	8.8	14.7	6.20335	92	
7	7	398	8.6	15.2	6.14707	93	

Figure 14.1 The TQM Datafile

Notes on the Datafile:

1. The variable *Month* is the number of months since the TQM program was introduced and is a number from 1 to 24.

2. The variable *Speed* is the number of items per minute that the machine is set to produce.

3. The variable *Waste* is the percent of manufactured product that is discarded during the manufacturing process.

4. The variable *RateOper* is the rate of operation and is a calculated number which measures the usable product throughput.

5. The variable *Delay* is a measure of the number of minutes per hour that the process is not operating.

6. The variable *Quality* is the percent non-defective product in those product lots that are shipped.

Load the datafile into Minitab and save it again under a different name.

Section 14.4 Regression and Correlation

While many people speak of regression analysis and correlation analysis as if they are one and the same, they do in fact have very different purposes. **Correlation** analysis is used to determine whether two variables are *related* in a linear manner and provides an estimate of ρ, the correlation coefficient. **Regression** analysis provides a model (equation) for predicting the value of a *dependent* variable using a set of one or more *independent* variables. In the simple case the model produced is linear, that is the equation of a straight line.

The manufacturing team for the floppy diskettes is interested in knowing how different variables in their process are related. Since they want to know *how* as well as *if* they should probably look at the data using regression analysis. Correlation information is usually provided along with a regression analysis, but not vice versa.

Simple linear regression assumes that there is a single, *controllable*, independent variable and a single, dependent variable and that these two variables are related according to the equation

$$Y = \beta_O + \beta_1 X.$$ where β_O is the y-intercept value and
β_1 is the slope coefficient.

The method of Least Squares is used to find estimates for the values β_O and β_1 and yields and equation of the form:

$$Y = b_O + b_1 X.$$

Section 14.5 Looking for Relationships

As always, before launching any statistical analysis you need to **look** at the data and assess what is going on. This will make the results of any analysis you do later on much more understandable. Since linear regression looks at the relationship between a controllable, independent variable and tries to find a model to predict values of the dependent variable, the first task is usually to decide which is which!.

Looking at the variables in the data set and knowing what the team are interested in finding out, it would appear that the ~~dependent~~ variable must be *Speed*, since this is the only variable that is within the control of the process. Suppose you look first at the relationship between *Speed* and *Waste*.

In order to determine whether there is even any reason to suspect that *Speed* and *Waste* are related you need to make a **Scatter Plot** of the data. You can do this in Minitab by making a **Plot** of the data. Unlike the plot you made to view a normal curve in Chapter 8, a scatter plot is a plot of the data points only, without lines connecting them.

From the main menu bar select **Graph** > **Plot**. In the **Plot** dialog box, position the cursor in the text box for **Y** and select *Waste* for the **Y** (dependent) variable and repeat to select *Speed* for the **X** (independent) variable. Under **Data Display** select **Symbol** for **Display**. After you click **OK** the graph shown in Figure 14.2 is displayed.

Scatter Plot of Waste vs. Machine Speed

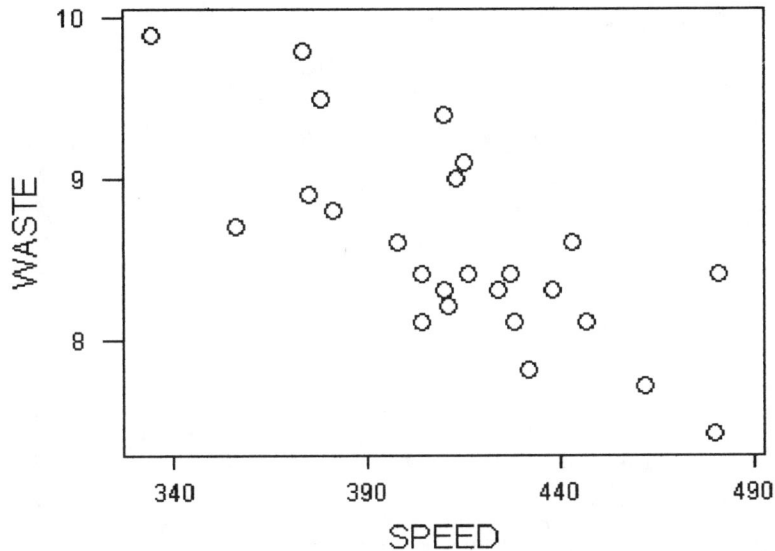

**Figure 14.2 Scatter Plot Showing Relationship
Between Waste and Machine Speed**

From the graph it appears that it is reasonable to assume that a relationship exists between the two variables and that it is indeed linear.

Section 14.6 Measuring Relationships

The first thing that you might want to do after looking at the graph it to determine the strength of the relationship between *Waste* and *Speed*. You can do this by finding the correlation coefficient for the two variables.

From the main menu bar select **Stat** > **Basic Statistics** > **Correlation**. The **Correlation** dialog box shown in Figure 14.3 will open.

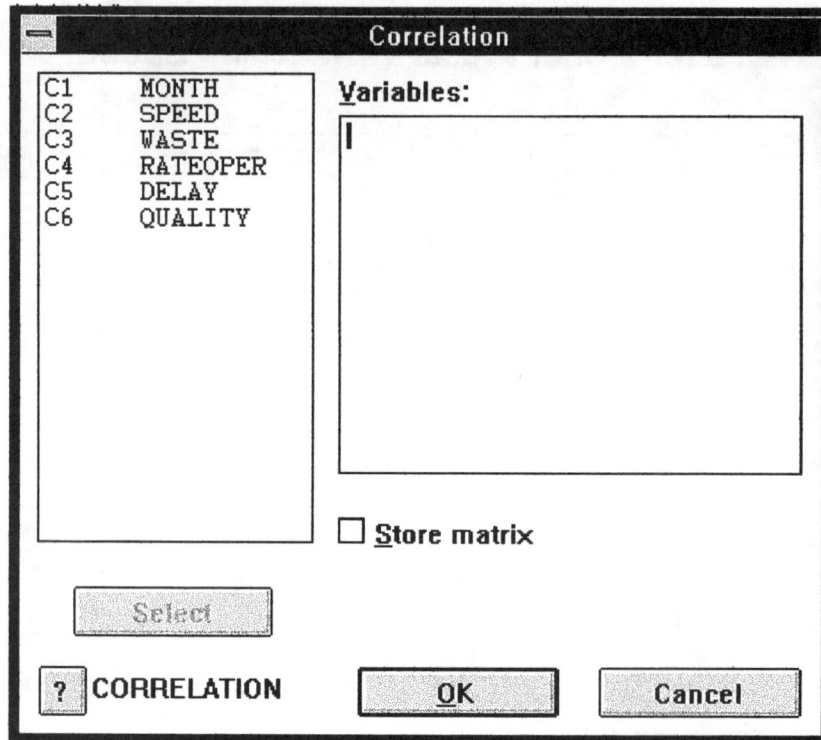

Figure 14.3 The Correlation Dialog Box

Position the cursor in the text box labeled **Variables:** and from the list, select *Speed* and *Waste*. Click on **OK** and the results shown in Figure 14.4 will appear in the **Session** window.

Notice that correlation analysis does not mention (or care about) dependence and independence. In correlation analysis, both of the variables can be random variables. This is NOT true in regression.

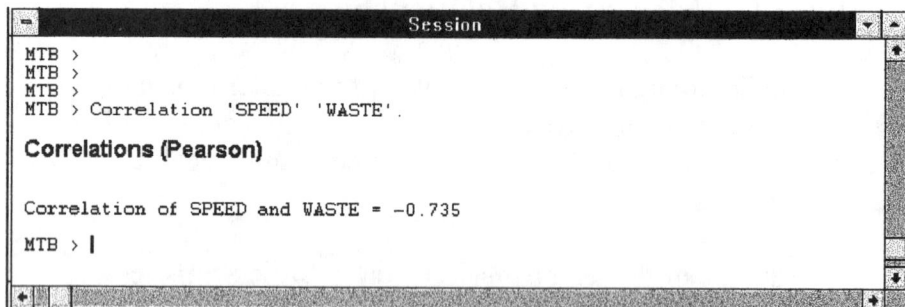

Figure 14.4   Correlation of Speed and Waste

The value of r, the sample correlation coefficient for *Speed* and *Waste* is -0.735. This indicates that there is a moderately strong relationship between the two variables and that they are related in such a way that when one variable increases, the other decreases. The process team thought that this would be true since **TQM** predicts that as speed and productivity increase, waste decreases.

Exercise 1. Do a correlation analysis for the variables *Waste* and *Delay*. Does there appear to be a correlation here? Does this make sense?

Section 14.7 Finding the Regression Model

The next step in the analysis is find the regression equation for *Waste* as a function of *Speed*. This will enable the team to predict what waste should be for different values of *Speed*. Such information can be used first to assess the **TQM** program and second to troubleshoot the process if *Waste* appears to be higher than the predicted value for a given speed.

From the main menu bar, select **Stat > Regression >Regression**. The regression dialog box shown in Figure 14.5 will open.

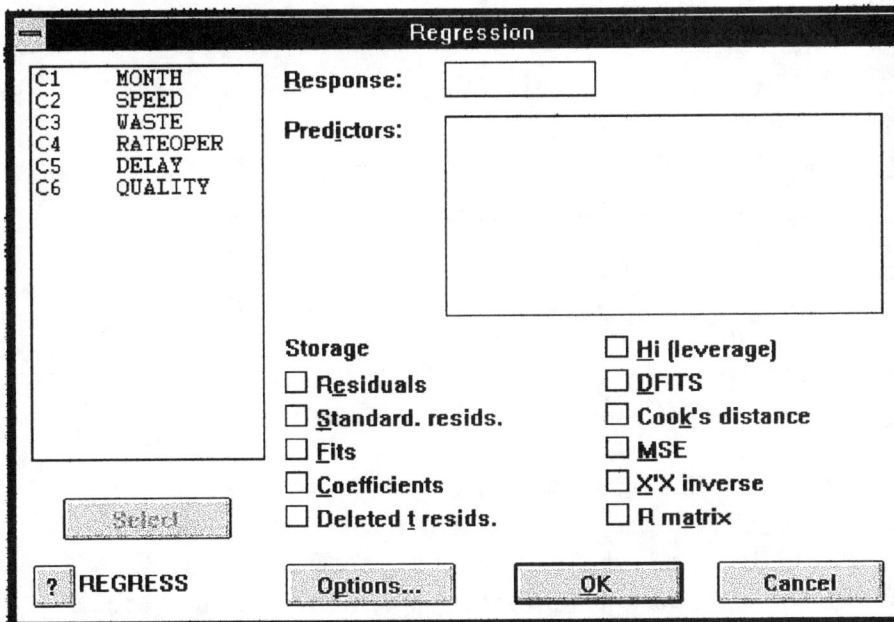

Figure 14.5 The Regression Dialog Box

There are a lot of different options associated with a regression. Right now, we are interested in obtaining the basic results of the procedure, the equation relating **Y** (independent) to **X** (dependent). Minitab refers to the **Y** variable as the **Response** variable and to **X** as the **Predictor** variable.

Position the cursor in the text box labeled **Response:** and select *Waste* from the list. Then move the cursor to the text box labeled **Predictor** and select *Speed* from the list. Click on **OK** and the results are displayed in the **Session** window. These results are shown in Figure 14.6.

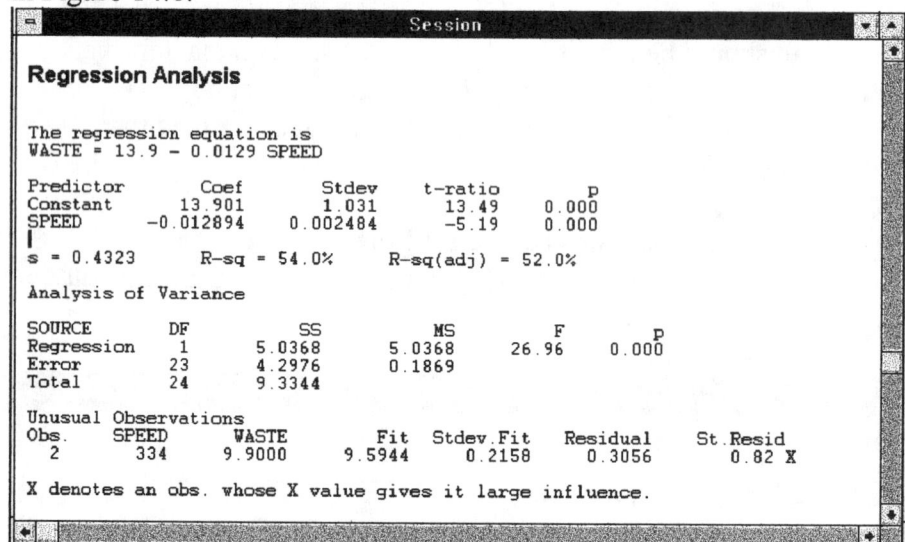

```
═                          Session                          ▼ ▲
                                                              ♦
Regression Analysis


The regression equation is
WASTE = 13.9 - 0.0129 SPEED

Predictor        Coef       Stdev      t-ratio        p
Constant       13.901       1.031        13.49    0.000
SPEED        -0.012894    0.002484       -5.19    0.000

s = 0.4323     R-sq = 54.0%     R-sq(adj) = 52.0%

Analysis of Variance

SOURCE        DF          SS          MS        F        p
Regression     1      5.0368      5.0368    26.96    0.000
Error         23      4.2976      0.1869
Total         24      9.3344

Unusual Observations
Obs.    SPEED     WASTE        Fit   Stdev.Fit   Residual   St.Resid
  2       334    9.9000     9.5944      0.2158     0.3056       0.82 X

X denotes an obs. whose X value gives it large influence.
```

Figure 14.6 Output from Regression Procedure

Even without selecting from the available options, the output gives a lot of information. The information of interest to us at the moment is at the top of the output. We see that the linear regression equation of Y (*Waste*) on X (*Speed*) is given by:

Waste = 13.9 - 0.0129 *Speed*.

Exercise 2. Change the value of *Waste* for the last data point from 7.4% to 12.0%. Redo the scatter plot and the regression. What happens to the model when there is a value which appears to be an outlier?

Note: Change the value back to 7.4 BEFORE you continue with this chapter

Exercise 3. Redo the correlation analysis. What happens to the value of the correlation coefficient?

Section 14.8 Testing Significance of the Model

The method of Least Squares, which is the method used to perform linear regression, will find a regression equation for any two variables. The fact that an equation is found does not in any way mean that the model is significant.

There are several ways to test whether a regression model is significant: 1) testing whether $\beta_1 \neq 0$; 2) testing whether the correlation coefficient, $\rho \neq 0$; and 3) testing the Regression Sum of Squares vs. Error Sum of Squares. When using Minitab, method 1 or method 3 is the easiest.

Method 1 tests to see if there is a relationship by testing whether the slope coefficient (β_1) is different from 0. You want to test the hypotheses:

*It is not sufficient to look at the **magnitude** of b_1 to make this decision, since the magnitude of the coefficient is related to the order of magnitude of the data.*

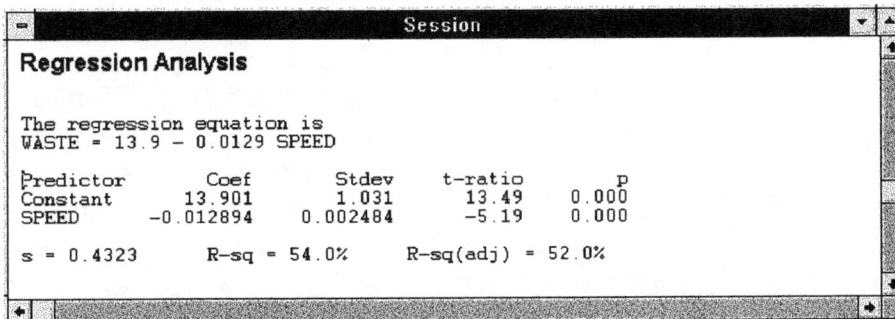

$$H_O: \quad \beta_1 = 0$$

$$H_A: \quad \beta_1 \neq 0$$

The top portion of the regression output is shown again in Figure 14.7.

```
 -                          Session                    ▼ ▲
                                                         ▲
Regression Analysis


The regression equation is
WASTE = 13.9 - 0.0129 SPEED

Predictor        Coef       Stdev      t-ratio         p
Constant       13.901       1.031        13.49     0.000
SPEED        -0.012894    0.002484        -5.19     0.000

s = 0.4323      R-sq = 54.0%     R-sq(adj) = 52.0%
                                                         ▼
 ◄                                                    ► ▲
```

Figure 14.7 Top Portion of Regression Output

The second part of the output gives the results of two different t tests about the regression coefficients. In addition to the hypotheses for the slope given previously you can perform a similar hypothesis test for the intercept (or constant) coefficient β_0.

From the output you can see that the p value for the slope (*Speed*) test is 0.000, which means that for any level of significance (α) greater than 0.000 (which is just about anything) you would reject H_O and conclude that the slope coefficient is not 0. That is, you conclude that there is a significant relationship between *Waste* and *Speed*.

The output also gives the value of R^2, the Coefficient of Determination. The value of R^2 can be viewed as the proportion of the variation in the dependent variable Y that is explained for or accounted for by the model (its relationship with X). For these two variables R^2 is 54% . Note that the value of R^2 is equal to the value of the correlation coefficient squared.

Section 14.9 Examining the Fit of the Model

Although the statistical test indicates that the regression model is significant, you may wonder just how good the model is at predicting. One way to find this out is to look at the differences between the observed value of Y and the predicted value of Y. These differences are known as the **residuals**.

From the menu bar select **Stat > Regression > Regression** to bring up the Regression dialog box. In addition to obtaining the regression relationship, Minitab allows you to calculate some other useful information and store it in columns of the worksheet. In this case you would like the predicted values of *Waste* for the values of *Speed* and the values of the residuals. From the section labeled **Storage:** click on the check boxes for **Residuals** and **Fits.** Click on **OK**. In addition to the regression output repeating in the **Sessions** window, you see that two new columns labeled *Fits* and *Resi1* are added to the worksheet as seen in Figure 14.8.

	C6	C7	C8	C9	C10	C11	C1
↓	QUALITY	FITS1	RESI1				
1	95	9.06572	-0.165720				
2	93	9.59439	0.305614				
3	94	9.31071	-0.610712				
4	93	9.02704	0.472963				
5	94	9.09151	0.708491				
6	92	8.98835	-0.188354				
7	93	8.76915	-0.169150				

**Figure 14.8 Worksheet With Columns for Fits
and Residuals**

If you look at the values in *Resi1* you see that the residuals, or errors, range from -0.610712 (predicted value higher than observed data) to 0.785580 (predicted value lower than observed data).

Another way to see the accuracy of the predictions is to look at a plot of the actual values of Y and the predicted values of Y on the same graph. From the main menu bar select **Graph> Plot**. For the first graph, input *Waste* for Y and *Speed* for X. For the second graph input *Fits* for Y and *Speed* for X. Then click on the arrow for **Frame**. The complete dialog box with the pop-up menu for frame is shown in Figure 14.9.

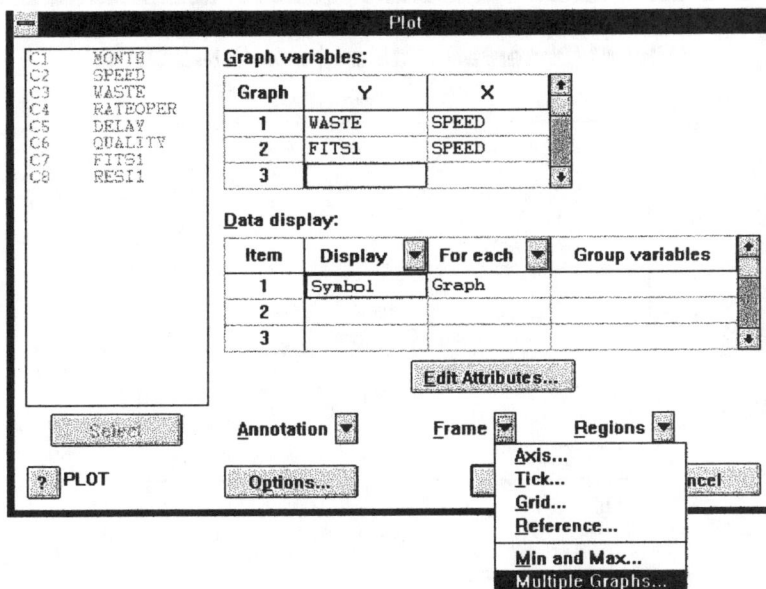

Figure 14.9 Dialog Box for Multiple Plot

From the **Frame** pop-up menu select **Multiple Graphs...** The dialog box shown in Figure 14.10 will open.

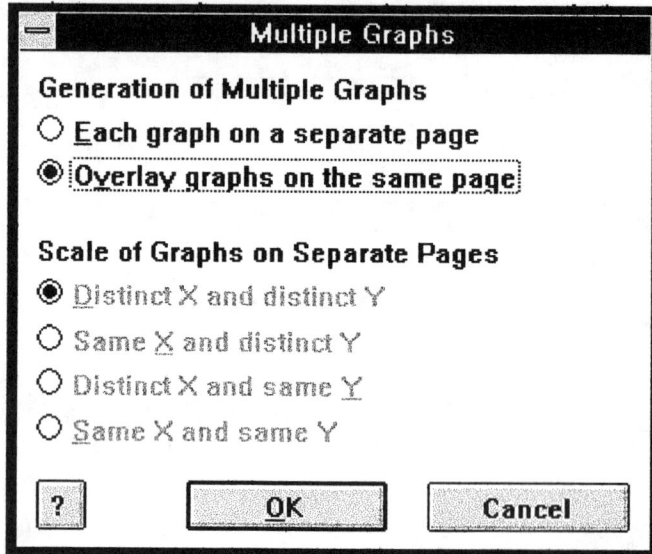

Figure 14.10 Multiple Graph Dialog Box

Click on the option button for **Overlay graphs on same page** and click **OK** twice to display the graph shown in Figure 14.11.

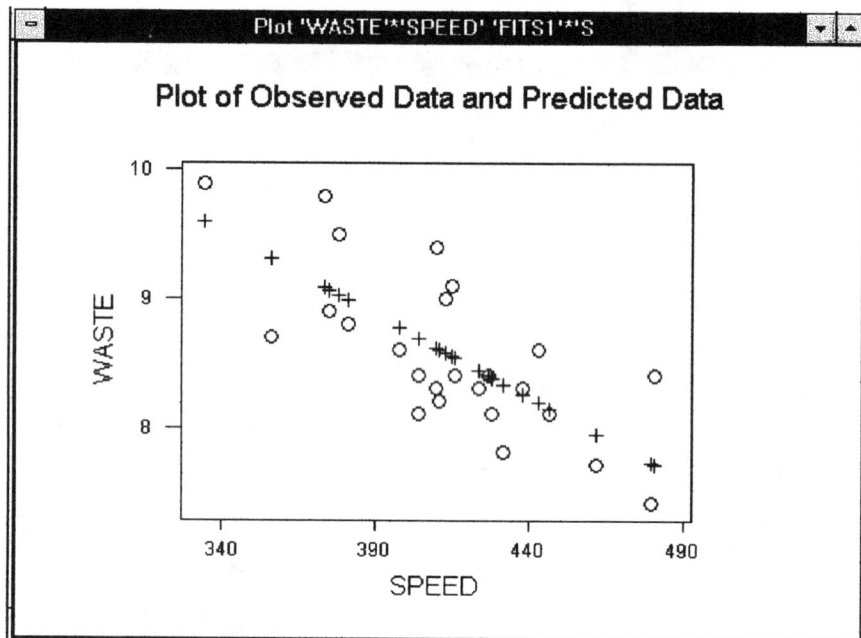

Figure 14.11 Plot of Actual and Predicted Values of Waste

Section 14.10 Interpolation and Extrapolation

Section 14.10.1 Finding Confidence and Prediction Intervals

Since the purpose of regression analysis is to provide a model for predicting the value of the dependent variable, you might want to look at the predictions for a model and decide how useful they really are.

Again, select **Stat** > **Regression** > **Regression** from the main menu bar. In the dialog box, enter only the variables to be used for X and Y. Then click on the command button labeled **Options...** The **Regression Options** dialog box shown in Figure 14.12 will open.

*NOTE: If the check boxes for **Fits** and **Residuals** are still checked, uncheck them by clicking in the check boxes.*

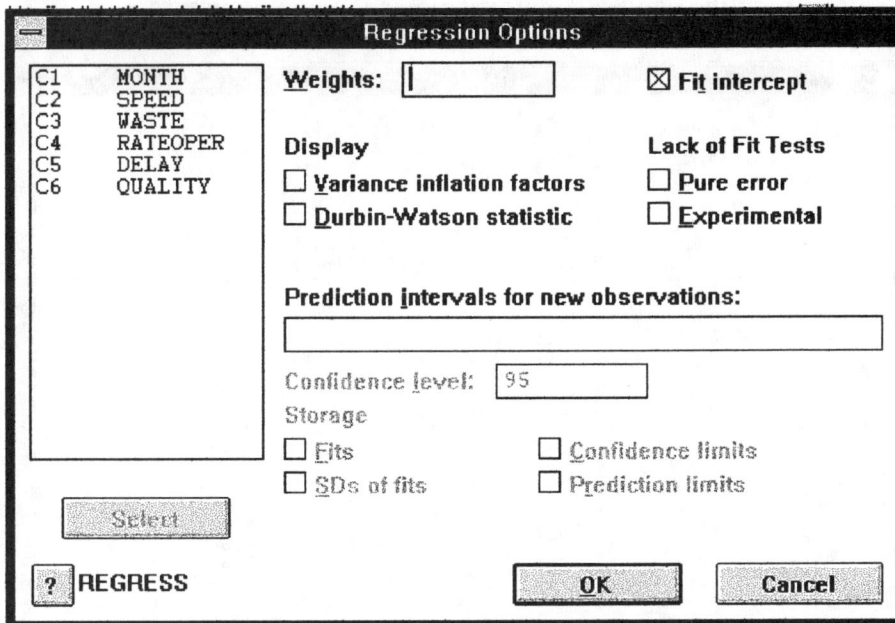

Figure 14.12 Regression Options Dialog Box

Minitab not only enables you to obtain the predicted values of Y for the selected values of X, but also provides $(1-\alpha)\%$ confidence intervals for the mean value of Y and $(1-\alpha)\%$ prediction intervals for individual values of Y for a given value of X. You may not have seen these intervals before since the formulas for computing them are rather complicated, but since you can easily obtain them in Minitab it is worth examining them.

It is possible to obtain prediction and confidence intervals for the entire set of X values, or for specific values. Since it is easier to obtain the information for the existing set of X values, we will do that first. In the text box labeled **Prediction intervals for new observations:** select *Speed* from the list box. This indicates that you want predictions for all of the values in the column *Speed*. Enter 95 (the default) for the **Confidence level:**. Click on the check boxes for **Confidence limits**, and **Prediction limits**. Click on **OK** twice to perform the procedure. The familiar output will appear in the **Session** window, with the additional information shown in Figure 14.13.

```
                                 Session
      Fit    Stdev.Fit        95.0% C.I.              95.0% P.I.
    9.0657      0.1291    (  8.7986,   9.3328)    (  8.1323,   9.9992)
    9.5944      0.2158    (  9.1479, 10.0408)     (  8.5947, 10.5940) X
    9.3107      0.1671    (  8.9649,   9.6566)    (  8.3518, 10.2697)
    9.0270      0.1237    (  8.7712,   9.2829)    (  8.0967,   9.9573)
    9.0915      0.1328    (  8.8167,   9.3663)    (  8.1558, 10.0272)
    8.9884      0.1184    (  8.7433,   9.2334)    (  8.0610,   9.9157)
    8.7692      0.0947    (  8.5731,   8.9652)    (  7.8535,   9.6848)
    8.6918      0.0897    (  8.5062,   8.8773)    (  7.7783,   9.6053)
    8.6144      0.0869    (  8.4346,   8.7943)    (  7.7021,   9.5267)
    8.5499      0.0865    (  8.3709,   8.7290)    (  7.6378,   9.4621)
    8.6144      0.0869    (  8.4346,   8.7943)    (  7.7021,   9.5267)
    8.5757      0.0865    (  8.3968,   8.7546)    (  7.6636,   9.4879)
    8.6015      0.0867    (  8.4221,   8.7809)    (  7.6893,   9.5138)
    8.6918      0.0897    (  8.5062,   8.8773)    (  7.7783,   9.6053)
    8.5371      0.0867    (  8.3577,   8.7164)    (  7.6248,   9.4493)
    8.3952      0.0926    (  8.2035,   8.5869)    (  7.4805,   9.3099)
    8.3823      0.0936    (  8.1887,   8.5759)    (  7.4672,   9.2975)
    8.4339      0.0902    (  8.2472,   8.6206)    (  7.5202,   9.3476)
    8.2534      0.1056    (  8.0349,   8.4718)    (  7.3327,   9.1741)
    8.3307      0.0978    (  8.1284,   8.5331)    (  7.4137,   9.2478)
    8.1889      0.1132    (  7.9548,   8.4231)    (  7.2643,   9.1135)
    8.1373      0.1198    (  7.8894,   8.3852)    (  7.2092,   9.0655)
    7.9439      0.1481    (  7.6375,   8.2503)    (  6.9985,   8.8894)
    7.6989      0.1884    (  7.3091,   8.0887)    (  6.7232,   8.6746)
    7.7118      0.1862    (  7.3266,   8.0971)    (  6.7379,   8.6857)
 X  denotes a row with X values away from the center
```

**Figure 14.13 95% Confidence and Prediction
Intervals for Speed**

The reason for deleting the columns for Fits and Residuals earlier and for making sure that the boxes were not checked again was so that we did not duplicate information in the worksheet.

In addition, the same information is added to the worksheet as shown in Figure 14.14.

	C7	C8	C9	C10	C11	C12	C1
↓	FITS1	RESI1	CLIM1	CLIM2	PLIM1	PLIM2	
1	9.06572	-0.165720	8.79861	9.3328	8.13227	9.9992	
2	9.59439	0.305614	9.14793	10.0408	8.59473	10.5940	
3	9.31071	-0.610712	8.96486	9.6566	8.35175	10.2697	
4	9.02704	0.472963	8.77117	9.2829	8.09674	9.9573	
5	9.09151	0.708491	8.81668	9.3663	8.15582	10.0272	
6	8.98835	-0.188354	8.74327	9.2334	8.06096	9.9157	
7	8.76915	-0.169150	8.57312	8.9652	7.85350	9.6848	

Figure 14.14 New Columns for Confidence and Prediction Intervals

Suppose you want to obtain the predicted values of *Waste* for a speed, say 400 PPM, which is not in the data set. In order to obtain the predictions you need to enter the value as a constants or put it in a new column. In this case we will enter it as a constant.

You can type Minitab commands in directly by typing them in the **Session** window. Position the cursor in the **Session** window, next to the MTB> prompt. To assign a value to a constant, at the prompt type:

let k1 = 400   and press ⏎.

Rerun the Regression procedure, making sure that you **uncheck** any check boxes that are checked from previous executions. The relevant part of the output is shown in Figure 14.15.

```
Unusual Observations
Obs.    SPEED       WASTE        Fit  Stdev.Fit    Residual     St.Resid
  2       334      9.9000     9.5944     0.2158      0.3056        0.82 X

X denotes an obs. whose X value gives it large influence.

     Fit  Stdev.Fit      95.0% C.I.           95.0% P.I.
  8.7434     0.0928   ( 8.5513,  8.9354)  ( 7.8286,  9.6582)

MTB > |
```

Figure 14.15 Prediction and Confidence Intervals for Speed = 400

Note: If you want prediction and confidence intervals of more than one value of the independent variable it is easier to enter them in a new column since this procedure only allows you to do one at a time.

From the output you see that for a *Speed* of 400 the predicted *Waste* is 8.7434%. If you look at the Confidence Interval for the mean value of *Waste* you see that it goes from 8.5513% to 8.9354%. This tells the team that when the machine is run at a speed of 400 PPM the *average* waste will range from 8.55% to 8.94%.

The values for the prediction interval range from 7.8286% to 9.6582%. This tells the team that if waste is measured on any given when the machine is run at 400 PPM the individual value of *Waste* will range from 7.83% to 9.66%. The confidence intervals and prediction intervals for a regression model give information on how *useful* the model is. If the intervals are too wide, because the relationship is not that strong, then the range on the intervals will be too wide to be meaningful or useful to the people who will use them. For example, if the prediction interval for a speed of 400 PPM were to range from 3.05% to 8.04% the manufacturing team would not find this useful in planning.

Exercise 4. Rerun the procedure using a confidence level of 99% What happens to the prediction and confidence intervals? What happens if you change it to 90%?

Using the model to predict values of Y is valid as long as the values you choose for X are within the range of the original data. That is, as long as you are **interpolating**. You might be interested in predicting *Waste* for speeds above 350 or below 480. This is called **extrapolation** and is **not advisable**. The model is only valid over the range of *Speed* observed!

Section 14.10.2 Plotting Confidence and Prediction Intervals

It is interesting to look at the relationship among the regression line, the confidence intervals and the prediction intervals. Minitab provides a mechanism for doing this. From the main menu bar select **Stat > Regression > Fitted Line Plot...**. The dialog box shown in Figure 14.16 will open.

Figure 14.16 Fitted Line Plot Dialog Box

This procedure also performs a regression analysis but the output is mainly graphical. Enter *Waste* for the **Response (Y):** and *Speed* for the **Predictor (X):**. Use the default (95%) confidence level and mark the check boxes for **Display confidence bands** and **Display prediction bands**. Enter a title and click on **OK**. The graph and associated information is shown in Figure 14.17.

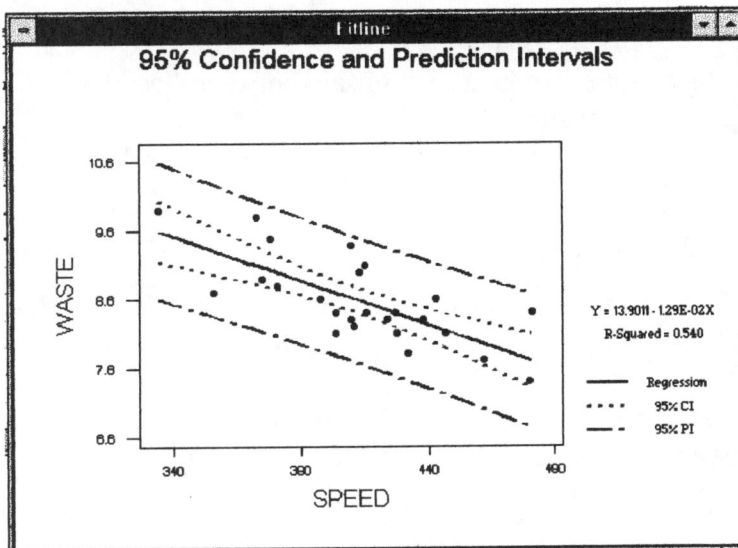

Figure 14.17 Output From Fitted Line Procedure

From the output you can see the relationship between the confidence interval and the prediction interval. Looking at the prediction interval you can also see that the confidence intervals are better (tighter) in the center (closest to $\overline{X}$) and get wider as you move away from the center. This is also true (but less obvious) for the predictions intervals.

Section 14.11 Investigative Exercises

In the following exercises you are asked to use the skills introduced in the previous chapters to extract information from the tissue strength datafile. You are provided with space to answer the questions and paste in graphical output from the program. If you do not have access to a printer, you can sketch the graphs in the spaces provided.

1. Create a scatter plot for *Speed* and *Delay*.

b) Does there appear to be a relationship between *Speed* and *Delay*? Describe the relationship.

c) Find the correlation coefficient for *Speed* and *Delay*. Does its value agree with your description of the relationship?

2. Find the regression model for *Delay* and *Speed*. Write the regression equation.

b) Is the regression significant?

c) Plot the regression line along with the data Does the plot indicate that the model does a good job of predicting *Delay?*

3. Use the model to predict *Delay* for speeds of 380 and 470. What do you learn from the prediction and confidence intervals?

b) Use the model to predict *Delay* for a speed of 600 PPM. Does the answer make sense? Do you think it is valid? Why or why not?

4. Use the Minitab software to perform a complete regression analysis for the variables *Speed* and *RateOper*. Is the model a good one?

5. Use the model you just found to predict the value of *RateOper* for a speed of 400. Do you think that the prediction and confidence intervals would be useful to the company?

6. Make a scatter plot of *Speed* and *Quality*. Do you think that there is a relationship between the two?

b) Perform a correlation analysis for the two variables. Does it indicate that there is a strong relationship?

7. Find the regression model for *Speed* and *Quality*.

8. Change the last 10 values for *Quality* to .97. Redo the regression analysis. Describe the results and how they compare to the model you found in the previous exercise.

9. The manufacturing company decides that it wants to know if there has been a steady increase in *RateOper* in the months since the TQM program was instituted. How might they accomplish this?

b) Plot *RateOper* against *Month*. Does it appear that there has been a steady increase?

c) Find a regression equation that predicts *RateOper* as a function of *Month*. What is the model?

d) Is it significant?

e) Find the residuals. Does the fit of actual vs. predicted appear to be good?

10. Prepare a report for management telling them about the effectiveness of the **TQM** program. Perform any additional analyses that you think you need to make a complete report. Include any plots that are relevant.

Chapter 15 "Do Your Tissues Rip?"

Pulling It All Together

Section 15.1 Overview

Statistical Objectives: There are no new statistical concepts to be learned in this chapter. The chapter is designed to allow you to use what you have learned through the exercises in this workbook. You will be asked to analyze a large data set but this time you will not be given any help. You will need to use statistical tools from several of the preceding chapters and it will be up to you to decide which tools are most appropriate. You are called upon to "pull it all together" !

Section 15.2 Problem Statement

Have you ever pulled the first tissue out of the box and had it tear? Have you ever opened a box of tissues and in trying to get one tissue out, ended up with several tissues? The problem causing both of these "nasty" things to happen is that there is not enough airspace in the box. This problem became an important issue for a large manufacturer of tissues when an unusually high number of complaints were registered.

Airspace is defined to be the amount of space between the top of the tissues and the top of the box. It is measured in millimeters and it should be at least 9 mm. Even if there is 9 mm of airspace when the box is manufactured, there may not be enough airspace by the time the customer opens the box. This is due to a phenomenon which is called "growback". As the box sits in the warehouse or on the supermarket shelf the tissues, which have been heavily compressed when they were put into the box, begin to expand or "growback". Thus the airspace is reduced.

In order to understand this dataset you need to know a little bit about how a box of tissues is made. In Figure 15.1 you see a picture of a hardroll. Each box of 250 tissues is made from 25 hardrolls each of which has different slit positions as shown in the diagram.

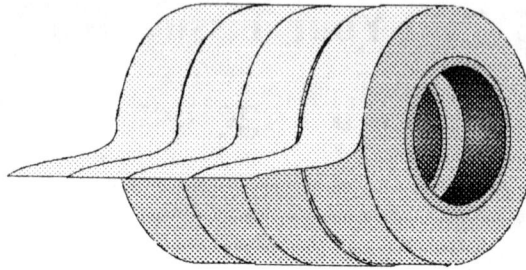

Figure 15.1 Schematic of a Hardroll

One tissue is taken from each slit and they are pressed into a tissue box and sealed.

An experiment was designed to investigate the hypothesis that the mean airspace differs by position in the hardroll from which the tissues were made. This hypothesis was proposed because of two facts:

1. As the hardroll sits in the warehouse, the outside collects moisture. Thus boxes made from tissues taken from the outside of the roll might have less "growback".

2. The core (very center) of the roll is very compressed (a hardroll weighs 1000 pounds). Thus boxes of tissues made from the core have the "life stretched out of them".

The data was collected in the following fashion:

• Sample 5 times through the hardroll. Since a hardroll lasts for 4 hours the 5 observations were taken at the beginning and once an hour thereafter. This resulted in observations at the top, the core and 3 in the middle.

• Each sample consisted of a case of tissues (24 cartons).

• A total of 4 cases were sampled every hour - one was used to obtain in-process measurements. The other 3 cases were saved for observations taken after 24 hours, 2 weeks and 4 weeks.

Section 15.3 Characteristics of the Data Set

FILENAME: CHAP15.MTW
SIZE   COLUMNS   3
            ROWS   480

The first seven rows of the actual datafile are shown in Figure 15.2

	C1	C2	C3	C4	C5	C6	C7
	Position	Time	Airspace				
1	▓	1	23				
2	1	1	25				
3	1	1	23				
4	1	1	23				
5	1	1	23				
6	1	1	23				
7	1	1	23				

Figure 15.2 First Seven Rows of CHAP15.MTW

Notes on the dataset:

1. The variable *Position* indicates from what part of the hardroll the sample was taken. The values for position are 1,2,3,4, or 5. Position 1 indicates it was taken from the outside of the roll and position 5 indicates it was taken from the core.

2. The variable *Time* indicates when the measurement was taken:
1= In-process, 2=after 24 hours, 3=after 2 weeks 4= after 4 weeks.

3. The variable *Airspace* is the measurement of the airspace measured in mm.

Read in the datafile named CHAP15.MTW. Remember to create a working copy of the datafile by resaving the file with a slightly different name.

Section 15.4 Investigative Exercises

Analyze the data using whatever techniques you feel are appropriate. Be sure to explain why you have selected a particular technique. Your analysis must address the following questions:

1. Is there a difference in airspace due to position in the hardroll?

2. Is there a difference in airspace by time tested?

3. How can you estimate "growback"?

NOTES

NOTES

NOTES

NOTES

NOTES

NOTES

NOTES

NOTES

NOTES

NOTES

NOTES

NOTES

NOTES